SPRINGER TRACTS IN MODERN PHYSICS

Ergebnisse
der exakten Natur-
wissenschaften

Volume **69**

Editor: G. Höhler

Associate Editor: E. A. Niekisch

Editorial Board: S. Flügge J. Hamilton F. Hund
H. Lehmann G. Leibfried W. Paul

Springer-Verlag Berlin Heidelberg GmbH 1973

Manuscripts for publication should be addressed to:

G. Höhler, Institut für Theoretische Kernphysik der Universität, 75 Karlsruhe 1, Postfach 6380

Proofs and all correspondence concerning papers in the process of publication should be addressed to:

E. A. Niekisch, Kernforschungsanlage Jülich, Institut für Technische Physik, 517 Jülich, Postfach 365

ISBN 978-3-662-15881-4   ISBN 978-3-540-46983-4 (eBook)
DOI 10.1007/978-3-540-46983-4

Astrophysics

Contents

On the Properties of Matter in Neutron Stars*

GERHARD BÖRNER

Contents

Abstract. A review of recent developments in the description of neutron star matter is presented, and its relevance to pulsar observations is discussed. Some aspects of the accretion of matter on neutron stars are reviewed, and some of the relevant properties of binary X-ray sources are presented. This review is aimed at the astrophysicist. For a detailed review of the nuclear physics involved see H. A. Bethe (1971) in Ann. Rev. Nuc. Sci., Vol. 21.

---

* This work was performed while the author held a NAS–NRS research-associateship at NASA/Goddard Space Flight Center.

1. Introduction

In 1968 the first source emitting a continuous train of radio pulses was discovered in Cambridge (Hewish et al., 1968). Subsequently more and more pulsars, as they were called, have been found up to a total of 64 at present. It turned out that their periods were kept very precisely, though not as precisely as those of atomic clocks. So one has for CP 1919, the first pulsar discovered, a period of

$$T = 1.337301101618 \, (7 \pm 10^{-13}) \, (\text{sec})$$

(see Manchester et al., 1972). The range of periods is rather large:

$$0.033 \, (\text{sec}) \leq T \leq 3.74 \, (\text{sec}) \, ,$$

and all periods seem to increase with time

$$0 \leq dT/dt \leq 4.23 \times 10^{-13} \, .$$

Let us first of all review briefly the elimination process (Maran and Cameron, 1969) by which one arrives at the conclusion that pulsars have to be rotating neutron stars:

Initially there were essentially two alternatives for explaining the observed periods of pulsars: Pulsations of very dense stars, with mean densities ϱ in the range of $10^8 - 10^9$ g/cc (white dwarfs, where $\sqrt{G\varrho} = 10^{0.5}$ to 10 rad/sec), discussed by a number of authors (cf. Cameron, Maran, 1969). Secondly rotating neutron stars with a frequency ω such that $\omega^2 \ll G\varrho$ (Gold, 1969; Pacini, 1968). The further possibility of dense contact binary systems, which leads essentially to the same relation between the period and the density as pulsation in the fundamental mode, was soon ruled out by the very high stability of the periods. This stability showed that there cannot be an emission of large amounts of gravitational radiation. The white dwarf pulsation hypothesis was ruled out consequently with the discovery of the fine structure of the pulses, with that of the regular rapid change of the angle of polarization during each pulse, and particularly with the discovery of the two young pulsars in supernova remnants with periods of less than 0.1 sec, which cannot be understood at all as stable pulsation of white dwarfs (Crab PSR 0531-21: $T = 33$ ms, Vela PSR 0833-45: 89 ms). Finally the slow secular increase of the periods typical for pulsars is another point in favor of the rotating neutron star hypothesis, because one would expect the loss of rotational energy to lead to a slowing down of the rotation. Rotating neutron stars also can account for the energy of the Crab Nebula (Finzi and Wolf, 1969; Wheeler, 1966) and in the Vela X remnant (Rees and Trimble, 1970; Börner and Cohen, 1971a). For all these reasons the model of a rotating neutron star as an explanation for the pulsar phenomenon has been generally accepted.

Although pulsars have become known only very recently, the concept of neutron stars goes back to the 1930's (Landau, 1932; Oppenheimer and Volkoff, 1939) when the equilibrium of a large body consisting of neutrons was considered. The temperature was assumed to be at absolute zero, and since the neutrons follow Fermi statistics, the pressure P of a gas of neutrons is related to the number density ϱ by

$$P = \tfrac{1}{5}(3\pi^2)^{2/3} h^2 m_n^{-8/3} \varrho^{5/3}, \qquad (m_n = \text{neutron mass}). \tag{1.1}$$

By integrating Einstein's equations for a spherically symmetric fluid

$$-dP/dr = G\frac{(m + 4\pi r^3 \, P/c^2)(\varrho + P/c^2)}{r^2(1 - 2G\, m/c^2 r)}, \tag{1.2}$$

$$m(r) = \int_0^r \varrho 4\pi r^2 dr \tag{1.3}$$

(r is a radial coordinate, such that the surface of a sphere of this radius is given by $4\pi r^2$) it was found that there exist stable configurations of neutrons, forming large bodies, so-called neutron stars.

That there must be a maximum stable mass becomes quite clear already from Newtonian considerations: The critical number of baryons A is reached when the addition of one more baryon (of mass m_n) will decrease the gravitational energy by an amount

$(-G m_n A/R)\, m_n \qquad (R = \text{radius of the body})$

larger than the gain in statistical energy

$dE/d\varrho = \tfrac{2}{3}h^2 m_n^{-8/3}(3\pi^2)^{2/3}\, \varrho^{-1/3}\, (\varrho\tfrac{4}{3}\pi R^3 = A).$

It turns out that $A \sim 10^{58}$.

Employing Einstein's general theory of relativity decreases the maximum mass, because the pressure gradient is increased on the right hand side of Eq. (1.2) compared to the Newtonian form of

$$dP/dr = -G m \varrho/r^2 \tag{1.4}$$

Oppenheimer and Volkoff find for their neutron gas star a maximum mass of

$M = 0.76\, m_\odot \ (m_\odot = \text{solar mass} \cong 2 \times 10^{33}\ \text{g}),$

a maximum central density of

$\varrho_c = 3.8 \times 10^{14}\ \text{g/cm}^3,$

and a maximum radius of

$$R = 9.42 \text{ km}.$$

This is the qualitatively correct picture of a neutron star: a massive, extremely dense and small object. More involved and more realistic equations of state change these parameters quantitatively, but the qualitative features remain the same.

The formation of such a small and dense object in a supernova event will also lead to a strong magnetic field. Because of the high electron number to be expected in such a star, the electrical conductivity will be very large, and the magnetic flux will be frozen in. Thus if we start with an object of radius $R = 10^{11}$ cm, $M = 1 M_\odot$, $\varrho = 1$ g/cm^3 and $B = 100$ gauss, we will end up with a neutron star of $R = 10^6$ cm, $\varrho = 10^{15}$ g/cm^3, $M = 1 M_\odot$, having a magnetic field of 10^{12} gauss.

These strong magnetic fields provide the link of communication between the rotating neutron star and the observer. All that is observable is the electromagnetic radiation, produced by charged particles accelerated in the strong magnetic fields around the pulsar and reaching us as continuous radiation or in pulsed form. No convincing model of how the pulses are formed has been put forward, but some gross features have been explained quite well. Thus it was shown by Goldreich and Julian (1969) that despite the strong gravitational attraction from a neutron star, there cannot be a vacuum outside the star (they took the magnetic field aligned with the rotation axis; the oblique rotator was treated similarly by Cohen and Toton, 1971).

Assume an interior magnetic field, which will be frozen in and which is consistent with an exterior dipole field. Because of the high conductivity of the neutron star interior, the condition that the electric field vanishes in the rest frame of the star is a good approximation. Consequently in the rest frame of an observer at rest at infinity, the electric field is given by

$$\boldsymbol{E} + \boldsymbol{V} \times \boldsymbol{B} = 0.$$

Via div $\boldsymbol{E} = \varrho/\varepsilon_0$, the charge-density associated with the electric field is given by

$$\varrho = -2 B_0 \omega \varepsilon_0 \cos \theta \qquad (\theta = \text{angle between } B \text{ and } \omega).$$

If it is assumed that the neutron star is surrounded by a vacuum, the solution of Maxwell's equations in the vacuum outside has to be matched to the interior solution via continuity of the magnetic field component normal to the surface and of the tangential component of the electrical field. It is found then that the quantity $\boldsymbol{E} \cdot \boldsymbol{B}$, which is zero inside the star, does not vanish outside. On the contrary

$$\boldsymbol{E} \cdot \boldsymbol{B} \sim R \omega B_0^2.$$

Thus near the surface charge layer of the neutron star the electric force along the magnetic field exceeds the gravitational force by a large factor of the order of 10^{13} for electrons and 10^{10} for protons. These ratios were obtained by using parameters typical of the Crab pulsar PSR 0531-21 ($B \sim 10^{12}$ gauss, $\omega \sim 200$ sec^{-1}, $R \sim 10$ km, $m = 1$ $m_\odot$). Thus if the surface region is ionized, the surface charge layer cannot be in dynamical equilibrium. A rotating magnetic neutron star must possess a magnetosphere, composed of charged particles traveling along the magnetic field lines. What we observe is the radiation from these charged particles injected into the magnetosphere.

No satisfying quantitative description of the electromagnetic link between the rotating neutron star and the radiation pattern of the pulsar has been given so far. Indeed, not even the case of a magnetosphere of radiating particles, where the axis of the magnetic field coincides with the rotation axis of the neutron star, has been solved. Whereas to explain the pulse producing mechanism one would have to treat the much more complicated case of at least a slight deviation from axial symmetry.

In the absence of a convincing pulsar mechanism theory the observations permit only a few rather crude conclusions on the physical properties of the rotating neutron star. For 22 of the 61 discovered pulsars, both frequency ω, and change of frequency $\dot{\omega} \equiv d\omega/dt$ have been measured. Then by determining their rate of loss of energy

$$\dot{E} = I\omega\dot{\omega},$$

we could in principle find the moment of inertia I of these neutron stars. This in turn would precisely fix mass and density profile of the star according to the equation of state used. Although the observations are not exact enough to permit definite conclusions in this line of investigation, certain limits on the physical parameters of a realistic neutron star can be derived. So as we proceed in the following sections to describe the physics of neutron star matter in the different density regimes, we shall always try to make clear how the different assumptions about the properties of matter affect the models of neutron stars and how these in turn relate to astrophysical observations. We hope then for a subsequent feedback of astronomical information on ideas about the fundamental structure of matter at high densities.

This paper is set up in the form of a review, trying to give an up-to-date survey of part of the work that has been done recently in this field. Although even for the last three years only an incomplete survey of the existing literature could be given, I have tried to incorporate the important ideas. Much of the material covered exists in the form of preprints or has been published quite recently. Several new ideas, speculations and criticisms of my own are incorporated too. These reflect the outcome of numerous inspiring discussions which I have had with many of my distinguished colleagues.

2. Qualitative Description of the Interior of a Neutron Star

2.1. Validity of the "Isotropic Fluid" Approximation

When we describe a star which contains superdense matter, we obtain the equation of state (pressure P as a function of density ϱ and temperature T), from the local physics (two particle interactions, etc.) without taking into account the gravitational field. We thus separate the influence of "global physics" – the gravitational field produced by this matter configuration via Einstein's equations – from our local physics, although the gravitationally induced binding energy in heavy neutron stars amounts to ~ 250 MeV per particle, or $\sim 25\%$ of the rest mass energy. So the influence of the gravitational field seems to be rather strong, and we must ask in how far the separation introduced to obtain nonrelativistically $P(\varrho)$ is a good approximation to reality.

Obviously the concept of deriving an equation of state nonrelativistically and plugging it into Einstein's equations would lose its validity if the gravitational potential varied strongly over distances of the order of

$$\hbar/m_n c \sim 10^{-13} \text{ cm}$$

($G =$ gravitational constant; $h =$ Planck's constant). Investigations of a system of fermions with gravitational interaction (Bondi, 1964; Bonazzola, Ruffini, 1969), which were treated in a certain approximation within the framework of general relativity, showed that the gravitational potential varies strongly over distances of 10^{-8} cm only if the density is well above 10^{42} g/cm³. But densities in the center of neutron stars are always less than 10^{16} g/cm³, and therefore one is on safe grounds when employing the usual procedure of finding $P(\varrho)$ from local physics and then putting it into Einstein's equations to determine global effects on the star.

2.2. Qualitative Picture of the Interior

A cross-section through a neutron star would approximately look like the picture drawn in Fig. 1, where the mass density of the star increases with depth.

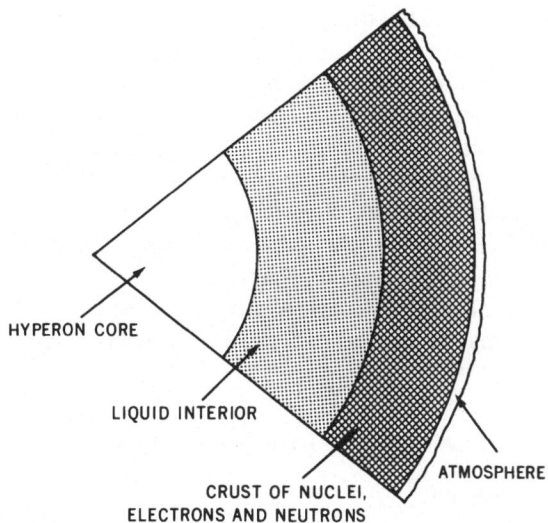

Fig. 1. The interior of a neutron star

In the different approaches for treating a system of nucleons with interaction there is general agreement on the qualitative features, despite some quantitative disagreement. As indicated in Fig. 1 there are several different states of matter (see e.g. Cameron's review article, 1970):

1. Under an atmosphere of a thickness of a few m (whose properties may be drastically influenced by the specific surface structure of the neutron star, see Section 3) we have.

2. a solid crust composed of neutron-rich nuclei arranged in a lattice and a degenerate electron gas. Up to a certain density ϱ_1 the charge Z and mass A of the nuclei is determined by equilibrium between β-decay and electron capture.

3. For densities above ϱ_1 unbound neutrons diffuse out of the nuclei ("neutron-drip"), and between ϱ_1 and a much higher density ϱ_2, we have equilibrium between free neutrons, neutron-rich nuclei in a lattice, and electrons.

4. For densities above ϱ_2 the nuclei disappear, and we have a small number (a few percent) of protons (and, of course, the same number of electrons) imbedded in the neutron sea. At a still higher density muons appear.

5. Finally at a density ϱ_3 and above new baryons probably make their appearance and hyperons like Σ^-, Λ^0, Ξ^-, Σ^0, Ξ^0,... etc. will be present together with neutrons, protons and electrons.

The uncertainties in the quantitative analysis throughout the different regimes 1 to 5 increase monotonically with density.

From computations of the cooling process of neutron stars (Cameron, Tsuruta, 1966) one finds that neutron stars cool very quickly from the initial, very hot state of formation down to about $10^8 \,^\circ$K in the interior (the surface will be cooler still, probably $\sim 10^5 \,^\circ$K (Cameron, 1970). The thermal kinetic energies of ~ 10 keV are therefore negligible compared to the several MeV per particle of nuclear energies or Fermi energies involved. In that respect, therefore, a neutron star is a very cold system. It is a good approximation to neglect the thermal energies and to consider the matter in neutron stars at absolute zero, $T = 0$.

3. Atmosphere and Surface of a Neutron Star

3.1. The Atmosphere

In the introduction we described the picture of Goldreich and Julian, where in the stationary state of a rotating magnetic neutron star ions and electrons continuously stream out along the field lines. Thus near the surface of the star there will be an atmosphere, which is thought to consist mainly of Fe^{56} – the endpoint of nuclear burning – and electrons (detailed models of the composition have been given by Rosen, 1968, 1969). The scale height in the atmosphere is determined from

$$dP/dr = -\varrho g,\tag{3.1}$$

where $g = GM/R^2$ is 10^{11} times the earth's g. If we neglect ionization the pressure is

$$P = \varrho\, kT/Am_n \quad (A = 56 \text{ for } Fe^{56},\, m_n = \text{neutron mass}),$$

so the scale height is of the order of a few cm. Hence the density increases rather rapidly, and finally with growing pressure the matter present will be fully ionized.

With increasing density the matter is gradually solidifying and continuously merging into a solid crust. In this region which corresponds to densities in the range $5 \leq \varrho \leq 10^4 \,\mathrm{g\,cm^{-3}}$, an equation of state due to Feynman et al. (1949) has usually been applied. In this approach a Thomas-Fermi model for atoms under pressure with a radius determined by the density is employed, but no effects of the magnetic field are considered.

Recent work (Ruderman, 1971; Mueller et al., 1971), however, has made it clear that because of the extremely strong magnetic fields near the surface of a neutron star, the gradient of the density becomes much steeper than one had previously thought. Also the general validity of the picture of a gaseous atmosphere as drawn above has become rather doubtful. This controversial point will be discussed in the following.

3.2. The Surface

Ruderman (1971) investigates the surface of a neutron star in the presence of a strong magnetic field. He finds that in huge magnetic fields ($B \geq 10^{12}$ gauss) matter forms a tightly bound, dense ($\sim 10^4$ g cm^{-3}) solid with the properties of a one-dimensional metal and a work function of the order of a keV. This model for atoms in strong magnetic fields is quite simple.

Assuming cylindrical symmetry around a uniform magnetic field B, one finds that the motion of charged particles perpendicular to the field is restrained. The particles can only move in certain quantized orbits (Landau orbitals) of radius

$$\varrho_n = (2n+1)^{1/2} \varrho \quad (n = 0, 1, 2, ...) \tag{3.2}$$

and

$$\varrho = (hc/eB)^{1/2} = 2.6 \times 10^{-4} B^{-1/2} \text{ cm} \tag{3.3}$$

The higher states are excited by integer multiples of

$$ehB/mc \sim 12 B_{12} \text{ keV}$$

(where B_{12} is the magnetic field in units of 10^{12} gauss). This high excitation energy assures that in the description of atoms in the stellar surface only the ground state is important. The lowest energy state of a single atom is realized by having the electrons in Landau orbitals which keep them (in directions perpendicular to B) much closer to their nuclei than the Coulomb field of the nucleus would by itself. The resulting atoms are shortened perpendicularly to B, and elongated along B. The lowest energy state of an assembly of such atoms is reached when they coalesce to form a tightly bound one-dimensional lattice parallel to B surrounded by a cylindrical electron sheath. This can be explained in two qualitatively different ways (the quantitative answer is the same in both cases!). Kaplan and Glasser (1972) consider an electron gas against a uniformly charged positive background in a strong magnetic field. When the Larmor radius for an electron eB/mc becomes smaller than the radius of the (spherical) volume per electron at the density considered, the system is essentially a dilute electron gas normal to the magnetic field and should undergo a "Wigner" transition to an ordered state, because then the quantum mechanical exchange correlation energy will dominate over the free-electron kinetic energy. This ordered state should resemble a two-dimensional lattice of charged rods with each rod behaving as a linear electron gas. The magnetic field B required for such a state is in the range of 2.5×10^{11} to 2.5×10^{13} gauss. Ruderman (1971), on the other hand, considers the energy of nuclei in a uniform density of electrons. The unscreened Coulomb repulsion between the nuclei is minimized by a bcc type lattice configuration.

A satisfactory quantum mechanical theory amalgamating these two points of view has yet to be done. We can, however, estimate the energy per atom from the classical Coulomb energy of a system of Z-charge nuclei sitting "like pearls on a string" (Ruderman, 1971) surrounded by a uniform density cylinder of electrons (radius $\sim 2l$). One finds

$$E_a = -[(Ze)^2/l] \times 1.2 , \tag{3.4}$$

where the lattice constant l is given as

$$l = 2.4 \, a_0 Z^{-1} \eta^{-4/5} \tag{3.5}$$

in a region (i) where

$$\eta = [(B/4.6) \times 10^9 \, Z^3]^{1/2} \gg 1$$

and

$$l \sim a_0 Z^{-1} \eta^{-12/15} \tag{3.6}$$

for

$$1 \gg \eta \gg Z^{-3/2} \quad [\text{region (ii)}];$$

a_0 is the Bohr radius.

E_a greatly exceeds the binding energy of an isolated atom [in region (i) and (ii)], and therefore Eq. (3.4), resp. (3.6), give the binding energy of an atom in the lattice. Thus iron nuclei ($Z = 26$) at the surface will be bound with an energy of -30 keV per atom, if we assume a magnetic field of 5×10^{12} gauss.

Adjacent chains will have strong Coulomb attraction when one is displaced half a lattice length along B relative to the other. An array of chains will then cohere so that the nuclei form a body centered orthorhombic lattice which is almost bcc. The mass density will be (Fe^{56}, $B = 5 \times 10^{12}$ gauss)

$$\varrho = 4 \times 10^4 \ \text{g cm}^{-3} .$$

The temperature of the surface will be $\sim 10^5 \, °K$ and evaporation of ions will be practically impossible. Hence there is a very sharp transition from a diffuse atmosphere, which contains probably mostly electrons, to the solid surface of the neutron star. This surface would then look at close examination rather similar to the skin of a not very smoothly shaved porcupine, with chains of Fe-atoms sticking out in all directions along the field lines.

Let us now discuss the consequences of Ruderman's (1971) model for pulsar observations. If ions cannot get out, then only electron currents would flow in the magnetosphere. The electrons are accelerated away from the surface by the electric field that is produced there (by the rotation of the magnetic field vector),

$$E = R\omega B/c \tag{3.7}$$

According to Goldreich and Julian (1969) particles stream out continuously from the surface of the neutron star, and a lower limit for the Crab pulsar ($\omega = 200$) seems to be 10^{33} particles/sec. If these particles are all electrons, they would in time build up an electric field opposite to the one given in (3.7). Thus the electric field that drags out charges from the neutron star will be weakened, then nulled, and no longer will charges of any kind flow out. The time scale can be estimated by simply looking at how long it would take for the surface charge of

$$N = (B\omega R/4\pi c) \, 4\pi R^2 \tag{3.8}$$

to be dissipated. Putting in numbers for the Crab pulsar we find

$$N/10^{33} = 0.03 \text{ sec}. \tag{3.9}$$

Thus the outflow of charges would cease after a time of less than 0.1 sec, and the pulsar would stop, if positively charged ions cannot get away from the surface.

Evidently Ruderman's (1971) description has very drastic consequences for pulsars. These consequences cannot be avoided by the assumption of an atmosphere composed of elements lighter than Fe^{56} which would not solidify by themselves (Mueller et al., 1971), because such an atmosphere will be transported away very quickly, and then the same problem as before has to be faced. If we do not want to invent a pulsar mechanism which is qualitatively different from the model proposed by Goldreich and Julian (1969), we have to find a way to get ions out of the surface.

Let us try: What energy do we gain in moving a positively charged ion over the lattice distance l? In region (i), i.e., using (3.5), we obtain from (3.7)

$$eE = 10^{-10} \, \omega(B/B_0)^{-1/5} \text{ ergs}, \tag{3.10}$$

where we introduced $B_0 = 5 \times 10^{12}$ gauss. This energy has at least to equal the binding energy, which in region (i) is

$$E_a = -10^{-7}(B/B_0)^{2/5} \text{ ergs}. \tag{3.11}$$

We find a lower limit for ω

$$\omega \gtrsim 10^3 (B/B_0)^{-1/5} \text{ sec}^{-1}. \tag{3.12}$$

This result indicates that for none of the known pulsars ($\omega_{max} = 190 \text{ sec}^{-1}$) particles can be extracted from its surface, a contradiction to the observations, which directly establish that at least the Crab ($\omega = 190$) and Vela ($\omega = 70$) pulsar send out a flow of charged particles.

If we look at region (ii), the situation gets slightly better, but the validity of this approximation remains dubious because here $\eta \approx 0.3$ and for (3.6) to hold we must have $\eta \ll 1$. This time we get

$$\omega \gtrsim 200(B/B_0)^{7/5} \tag{3.13}$$

So the Crab pulsar barely makes it. One can, of course, juggle around with B and try by variation of B to bring ω closer to realistic values. But in (3.12) ω does not vary strongly with B, and in (3.13) B would have to be as low as 10^{11} gauss, if we want to incorporate the slowest pulsars. This would then invalidate Ruderman's considerations anyhow. Then it would seem that Ruderman's concept cannot be reconciled to the qualitative picture of a pulsar as proposed by Goldreich and Julian.

There is, however, still another method to extract ions from the neutron star. The electric field exerts a pressure $E^2/4\pi$ on each atom, pulling at the interface of 2 atoms in the chain, and if this force is big enough to break the chain at any one point, the whole piece of the chain will be lifted up from the surface of the neutron star. The energy gained by lifting such a piece of a chain of atoms over one lattice distance is

$$(E^2/4\pi) \, l^2 \cdot l, \tag{3.14}$$

and analogously to the foregoing considerations we find for ω in region (i)

$$\omega \gtrsim 200(B/B_0)^{-1/5} \tag{3.15}$$

and in region (ii)

$$\omega \gtrsim 30(B/B_0)^{7/5}. \tag{3.16}$$

Again for region (i) there is, even through strong variations of B, no way to bring ω down to 2 (PSR 0525 + 21 has $\omega = 1.7$). In region (ii) the situation is more favorable; even without varying B the three fastest pulsars have the ω required by (3.16). When we let B go as low as

7×10^{11} gauss, then all the pulsars observed so far could work in the currently accepted way and have a solid surface of the kind predicted by Ruderman (1971). However, estimates of the magnetic field in pulsars give field strengths around 3×10^{12} gauss (assuming dipole fields), and one would expect the actual fields near the neutron star surface to be still larger. Furthermore, there does not seem to be a correlation between magnetic field strength and period of the pulsar, as we would predict it here.

I would like to point out that it is, indeed, still an unsolved problem to reconcile Ruderman and Goldreich and Julian, especially in view of the fact that the approximation of region (ii) is probably not too reliable, and if one relies on the approximation of region (i) the discrepancies are evident.

It may nevertheless be interesting to point out that in the model developed above, electrons would flow out from the pulsar as single charges, whereas the ions would come out in little chunks, each piece consisting of 10 to a few 100 Fe^{56} atoms.

4. Nuclear and Solid State Physics in the Crust

4.1. The Range of the Densities Below the Neutron Drip Line

Below the surface densities are immediately greater than $10^4 \, \text{g cm}^{-3}$, and the nonrelativistic Fermi energies of electrons increase quickly beyond 10 keV. Since we expect temperatures in these regions of less than $10^8 \, ^\circ$K, the electrons form a degenerate plasma. This makes the star optically thick, because photons (w, k) can only propagate if

$$w^2 = c^2 k^2 + w_p^2, \tag{4.1}$$

where the plasma frequency

$$w_p^2 = 4\pi n_e e^2 / m_e \tag{4.2}$$

increases as $\varrho^{1/2}$. Already for $\varrho \geq 2 \times 10^5 \, \text{g cm}^{-3}$ one has that

$$\hbar w_p / k > 10^8 \, ^\circ \text{K}$$

and photons can no longer be produced by thermal excitation. So we will find no photons inside a neutron star except in the outermost few meters.

At a density of $\varrho \geq 10^7 \, \text{g cm}^{-3}$ the electrons become completely relativistic. Nuclei will no longer be screened by clouds of electrons, but rather the negative charges will form a uniform background. The nuclei will then feel their relative Coulomb charges, repel each other, and in trying to minimize their energy, arrange themselves in a lattice, probably of bcc type. So the surface (almost bcc) lattice induced by the magnetic field will be replaced by a Coulomb lattice here. The lattice energy $\sim Z^{2/3} e^2/hc$ is negligible with regard to the energy balance at lower densities, but becomes important in determining the most stable nucleus.

At still higher densities above $8 \times 10^6 \, \text{g cm}^{-3}$ the electron capture process

$$p + e^- \rightarrow n + v \tag{4.3}$$

becomes energetically more favorable then the inverse reaction, the β-decay of the neutron. The high Fermi levels of the electrons make the neutron into a stable particle. New equilibrium configurations turn up where Fe^{56} is no longer the most stable nucleus, and more and more neutron-rich nuclei appear. These nuclei would be unstable under laboratory conditions, but here they are stabilized by the high Fermi levels of the electrons.

Recently Baym et al.(1971a) re-determined the most stable nucleus present at a given density under these conditions (Table 1). A more detailed treatment of this region, however, would have to take into account the way in which the species of nuclei was determined at the time it was frozen into the lattice during the cooling down of the initially very hot neutron star (the temperature of the very young star was $\sim 10^{11}$ to $10^{10}\,°\text{K}$, enough to melt the lattices considered here). Probably the most stable nucleus was not always realized, and certain defects might be in the lattice giving rise to creep phenomena. Sophisticated solid state physics would have to be employed in describing the crust.

Table 1. Nuclei in equilibrium with electron fermi gas (from Baym et al., 1971a)

Density (g/cm^3)	Equilibrium nuclei
7.8 E 6	$_{26}Fe^{56}$
2.8 E 8	$_{28}Ni^{62}$
1.2 E 9	$_{28}Ni^{64}$
8.0 E 9	$_{34}Se^{84}$
2.2 E 10	$_{38}Zn^{80}$
1.1 E 11	$_{38}Zn^{82}$
1.8 E 11	$_{42}Mo^{124}$
2.7 E 11	$_{40}Zr^{122}$

Already, however, the simplified picture given by Bethe et al. (1970) can explain some basic physical properties quite adequately, and we therefore briefly report on that paper. The total energy is a sum of the energy of nucleons in nuclei E_N, the free electron energy, and the lattice energy. The energy per nucleon E_N/A is given by the semiempirical mass-formula (Myers, Swiatecki, 1968)

$$E_N/A = -c_1 + c_2 Z^2 A^{-4/3} + c_3 (N-Z)^2 A^{-2} + c_4 A^{-1/3} \qquad (4.4)$$

(N: neutron number; Z: proton number, $A = N + Z$). Secondly for stable nuclei we have an equilibrium between β-decay and electron capture:

$$E_p + E_e = E_n + (m_n - m_p) c^2 \qquad (4.5)$$

[subscript (p, e, n) for (proton, electron, neutron) resp.].

The Fermi energy, or the chemical potential, as the energy of the highest occupied state, is given by

$$\mu_n = \partial E_N/\partial N, \qquad \mu_p = \partial E_N/\partial Z \qquad (4.6)$$

for neutrons and protons respectively. Equation (4.5) must also hold for the Fermi energies

$$\mu_n - \mu_p = \mu_e. \qquad (4.7)$$

The small neutron-proton mass difference is omitted. It does not influence the results appreciably because μ_e will turn out to be about 20 MeV or more. At higher densities the electrons are a highly relativistic Fermi gas, and therefore

$$\mu_e = c \varrho_N^{1/3} x^{1/3} \qquad (4.8)$$

where $x = Z/A$, ϱ_N = number density of nucleons bound in nuclei; c = a constant.

When we now determine the minimum of E (neglecting the small lattice energy

$$E_L \sim -2Z^2 e^2/a, a^3 \varrho_N = 2),$$

we find all quantities as functions of x:

$$A = (c_4/2c_2) x^{-2}, \qquad (4.9)$$

a remarkably simple formula. A increases with decreasing x (increasing density); even $Z = Ax$ increases. The neutron Fermi energy is a monotonically decreasing function of x, for $x > 0.04$. μ_n is zero for $x = 0.32$, while the total energy at that point is still negative $E/A = -1.6$ MeV. For x lower than 0.32, μ_n is positive; therefore, free neutrons appear and matter consists no longer only of nuclei, but of nuclei immersed in neutron matter. The density ϱ_1 of this neutron drip line was found by Bethe et al. (1970) to be

$$\varrho_1 = 2.8 \times 10^{11} \text{ g cm}^{-3} \tag{4.10}$$

in agreement with other authors (Cohen et al., 1969). Baym et al. (1971 b) using a slightly different mass formula find $\varrho_1 = 4 \times 10^{11} \text{ g cm}^{-3}$.

4.2. Neutron Drip Line to Break-Up of Nuclei

The number-density of free neutrons increases very rapidly with density, whereas the density of protons and electrons does not change rapidly at first and is then always a few percent of the neutron density in this region. Therefore the revised approach of Leung and Wang (1971) to treat the nuclear matter in this region as consisting purely of neutrons (neutron matter) seems reasonable.

In reality one has to deal with a system of electrons, free neutrons, protons and neutrons bound in nuclei. This is a typical case of nuclear matter and can best be treated by the many-body methods of the Brueckner-Bethe-Goldstone theory (see e.g. Day, 1967). This theory is essentially a sophisticated perturbation technique, adapted to many-body problems. Starting from a two-particle interaction described by a potential the two-particle correlations are computed. The ultimate aim is to describe real nuclei; but so far one has just been working on the reproduction of the properties of nuclear matter ($N = Z$) of infinite extension. In equilibrium (as one knows from large nuclei) the binding energy should be -16 MeV, the average particle distance $r_0 = 1.12$ fm, and the Fermi momentum $k_F(\varrho = \frac{2}{3}\pi^{-2}k_F^3) = 1.36$ fm^{-1}. The potential used to describe the interaction of two nucleons is – in most of the cases discussed here – the Reid soft-core potential (Reid, 1968), which fits scattering data very well, and gives reasonably good values for the binding energy of nuclear matter: -9 MeV in the approach of Nemeth and Sprung (1968), -11 MeV by the improved computational methods of Siemens (1970). To reproduce the exact value of -16 MeV corrections are introduced in all interactions of isospin $T = 0$ and $T = 1$, as in approximation (1a) of Nemeth and Sprung (1968), where the potential energy of all two-body states is multiplied by a factor of 1.22, or just in the $T = 0$ channel, as e.g. in (1b) of Nemeth and Sprung (1968) or Siemens (1970). The second approximation seems to be rather more accepted at present, since many authors feel that the discrepancy lies mainly in the tensor force in the $^3S_1 - ^3D_1$ state ($T = 0$ interaction). Matter consisting mainly or purely of neutrons should then be well described by using the nuclear matter calculations with the $T = 0$ channel switched off.

It should be stressed here that the attempts to find agreement between theory and experiment in the nuclear matter many-body calculations are meaningful only when a reliable method of computation has been established. One has to show explicitly that higher order terms, that have been neglected in a two-body correlation computation, contribute much less than the terms considered. Near the saturation density of $\varrho = 2 \times 10^{14}\,\mathrm{g\,cm^{-3}}$ it has been shown mainly through the work of Bethe that in the framework of the BBG method the third-order correlations are indeed small (Rajaraman, Bethe, 1967). This has not been done in any other scheme, and, therefore, the BBG method is to date the one reliable method for doing nuclear matter computations. In Fig. 2 the results of several such neutron matter computations for the equation of state can be seen [curves (2), (3), (5), (6)]. They all agree remarkably well in the region of $10^{12}\,\mathrm{g\,cm^{-3}}$ to $10^{14}\,\mathrm{g\,cm^{-3}}$.

The clustering of some nucleons into nuclei does not change the pressure-density relation appreciably in this region (as e.g. shown in Börner and Sato, 1971). The question of the equilibrium species of

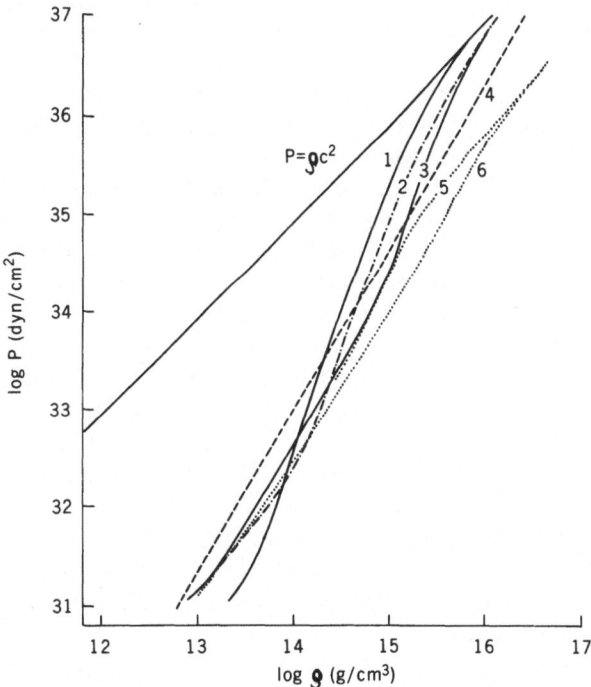

Fig. 2. Pressure as a function of density. 1. Cohen et al. (1969) – CCLR; 2. Bethe et al. (1970), Pandharipande (1971) – BBS; 3. Baym et al. (1971a) – BPS; 4. Free neutron gas; 5. Leung and Wang (1971) Model (II), 6. Leung and Wang (1971) Model (I)

nuclei coexisting with neutron matter has, however, great interest in itself. Also the lattice formed by nuclei in the outer layers of a neutron star determines the elastic properties of that region. This problem was first treated by Cohen et al. (1969), who took the Levinger-Simmons (1961) potential with constants adjusted by Weiss and Cameron (1969) to fit certain nuclear matter results. They do not use many-body techniques but calculate nuclear matter energies by simply taking first order expectation values of the energy (for a detailed discussion see Bethe et al., 1970).

Bethe et al. (1970) treated the problem again and employed calculations by Nemeth and Sprung (1968) carried out using full nuclear matter theory and the Reid (1968) soft core potential.

Stable nuclei can exist simultaneously with neutron matter only if the Fermi energies of the neutrons are the same in the nuclei (μ_{nN}) and in the neutron gas (μ_{nG}):

$$\mu_{nN} = \mu_{nG} \tag{4.11}$$

If this equality were not true, e.g., if $\mu_{nN} > \mu_{nG}$, then neutrons would evaporate from the nuclei until (4.11) is fulfilled. Since $\mu_{nG} > 0$, one finds that μ_{nN} must also be positive. This was already realized by Harrison et al. (1964).

Nuclei with $\mu_{nN} > 0$ are unfamiliar; however, they are easily interpreted physically. μ_{nN} is the largest energy of any neutron in the nucleus, but it is still only a few MeV, whereas the kinetic energy of the same neutron in the nucleus is still about 30 MeV. Therefore for most of the neutrons in the nucleus, the total energy ε is still negative; they are bound in the usual sense. A second class of neutrons will have energy $\varepsilon > 0$, but high angular momentum ($l > 4$); these neutrons feel a potential barrier preventing them from leaving the nucleus in spite of their positive energy; they may thus still be considered as essentially bound. A third kind of neutron with $\varepsilon > 0$ and low angular momentum will essentially be able to go freely between nucleus and neutron gas. Thus for $\mu_{nN} > 0$ the nuclei have the somewhat unusual property that some neutrons can pass freely in and out, but the nucleus still forms a compact structure in the surrounding uniform neutron gas.

The electron Fermi energy is again given by the formula for the ultrarelativistic Fermi gas, and in simple neutron matter (i.e. without imbedded nuclei) there must be β-equilibrium, hence the proton Fermi energy in neutron matter is

$$\mu_{pG} = \mu_{nG} - \mu_{eG} . \tag{4.12}$$

In fact, usually the concentration of protons in neutron matter is calculated by determining the proton Fermi energy which is mainly due to the potential energy of a proton in the neutron gas and is therefore strongly negative; then the electron density must be chosen so that (4.12) is fulfilled.

In a mixture of neutron matter and nuclei the density of protons in the neutron gas phase is zero and hence the proton Fermi energy is less than for neutron matter of the same neutron density. However, this difference is very small; it is due to the proton kinetic energy

$$T_p = 2\mu_e/2Mpc^2 .\tag{4.13}$$

If one takes into account the Coulomb energy of the lattice of nuclei, this difference again almost cancels out (Baym et al., 1971 b).

It was found by Bethe et al. (1970) that at a density of $\varrho \sim 5 \times 10^{13}$ g cm^{-3} the proton Fermi energy in the gas ($\mu_{pG} = -58$ MeV) decreased below that of protons in the nuclei. Thus at that density there was a sharp transition, where protons bound in nuclei distributed themselves in the surrounding neutron gas, dissolving the nuclei. Bethe et al. (1970) neglected both the proton kinetic energy and the lattice energy of nuclei. Langer et al. (1969) took the proton kinetic energy into account and found the transition to occur more smoothly but also in a narrow region around the same density. Including also the lattice energy again gives a sharp transition (Baym et al., 1971 b).

In both cases the nuclei were described by a semiempirical mass formula, and the dependence of the surface symmetry energy and the Coulomb energy term on the density of the neutrons outside was neglected (Cohen et al., 1969) or underestimated (Bethe et al., 1970).

Baym et al. (1971 b) carried on this earlier work and found that to determine the nuclear size the Coulomb energy of the lattice formed by the nuclei becomes more and more important. They changed some terms in the semi-empirical mass formula to obtain better fits to measured nuclear radii – which is a dangerous thing to do – and they included besides the Coulomb lattice energy the dependence on the outside neutron density of the Coulomb energy, which tends to favor small A nuclei, and of the nuclear surface energy, which tends to favor nuclei with large A. They use the improved neutron matter calculations of Siemens (1970) and Siemens and Pandharipande (1971). They find that as the density increases the nuclei fill more and more of space. Finally at a density of 2×10^{14} g cm^{-3} they begin to touch. With further increase in density the nuclei disappear discontinuously in a first order phase transition around 3×10^{14} g cm^{-3}. At that point, as can be seen from Table 2, they obtain nuclei with $Z \sim 200$, $A \sim 2500$.

Barkat and Buchler (1971), on the other hand, base their approach on a Thomas-Fermi model for nuclei and on BBG nuclear matter calculations and find the proton number Z to be relatively small

Table 2. Mass number A and charge Z of nuclei in neutron matter as a function of matter density ϱ. BBS: Bethe et al. (1970); BBP:Baym et al. (1971 b); BB: Barkat and Buchler (1971)

ϱ	BBS		BBP		BB	
g/cm³	Z	A	Z	A	Z	A
2.8 E 11	39	122	40	122	34	80
7.8 E 12	45	159	49	178	35	108
1.5 E 13	47	177	54	211	34	
2.2 E 13	49	190	58	241	34	
3.8 E 13	51	205	67	311	32	
5.1 E 13	51	211	74	375	30	
7.7 E 13			88	529	27	
1.3 E 14			117	947		
2.0 E 14			201	2500		

($Z \sim 33$ to 36). They do not determine the much more uncertain values of A. In agreement with Baym et al. (1971 b) they find that the nuclei disappear by way of a first order phase transition around normal nuclear density $\varrho_2 = 3 \times 10^{14}$ g cm^{-3} (both approaches find contrary to Bethe et al. (1970) that always $\mu_{pG} > \mu_{pN}$).

Similar results have been obtained by El Eid (1973) using a droplet model, and by Ravenhall et al. (1971), who payed special attention to the surface energy and treated semi-infinite nuclear matter in Hartree-Fock approximation. Recently Negele (1973) has carried out Hartree-calculations, and finds that the nuclei are similar to those found by the authors quoted above, with $Z \leqq 40$ throughout. He finds that due to shell effects Z changes discontinuously with increasing density.

The properties of the equation of state are, however, not appreciably affected by these different results obtained for the shape and size of the nuclei. The nucleon-nucleon interaction is the most important feature. As can be seen from Fig. 2, for densities up to normal nuclear density all the equations of state that are obtained via realistic nuclear matter calculation, where one reliable and well-tested method is the BBG method, agree rather well. Even Leung and Wang (1971) who treat the case of pure neutron matter, with no protons or other particles, do not depart from this general picture. Other equations of state obtained with the same methods but different potentials agree rather well too.

4.3. Superfluidity

In the range below normal nuclear densities, the neutrons lying at the top of the Fermi sea have an attractive potential for one another. This attractive force between neutrons arises from the $^1S_0$ interaction, which changes its phase shift from attractive to repulsive at normal nuclear densities. Thus below $\varrho_2 = 3 \times 10^{14}\,\mathrm{g\,cm^{-3}}$ pairs of neutrons will form, very similar to electron pairs forming a superconductor. One expects that the free neutrons in the crust will probably form a superfluid, because of this mutual attraction between pairs of particles. Similarly the protons will form a superfluid at about the same densities [see Cameron (1970) for extensive references].

In the spectrum of single-particle energy states immediately above the Fermi surface an energy gap of about 1 MeV will form in the superfluid. Clark and Yang (1971) basing their computations on the Bogoliubov-Valatin transform find a gap energy of 2.3 MeV at a density of $6 \times 10^{13}\,\mathrm{g\,cm^{-3}}$. Thus the gap energy is of the same order as the neutron Fermi energy and it may be that pairing effects can considerably influence the properties of neutron matter at those densities.

It has been pointed out that the rotation of the neutron star imposes interesting conditions on the superfluids. The superfluid will normally not partake in the rotation, but quantized vortex lines will be established throughout the interior. It is expected that the separation between quantized vortex lines will be $\sim 10^{-2}\,\mathrm{cm}$, and the radius of a core of a vortex line $\sim 10^{-12}\,\mathrm{cm}$ (Baym et al., 1969). Whereas the superfluid by itself could not be expected to contribute to the moment of inertia of the star, the mixture of superfluid and quantized vortex lines will probably rotate as a rigid body.

The crust, with a Coulomb lattice of nuclei through which the superfluid neutrons move, presents many interesting problems to the solid state physicist. There are indications that some of the properties of the crust manifest themselves in events that can be observed in pulsars (see Section 4.4).

At normal nuclear density the $^1S_0$ phaseshift changes sign, the neutron-neutron $^1S_0$ interaction becomes repulsive, and this type of superfluidity will disappear.

But then, above $2 \times 10^{14}\,\mathrm{g\,cm^{-3}}$ a significant attraction between pairs appears in the $^3P_2$ state. Thus a new type of superfluidity would be present up to very high densities. However, this superfluid state would be anisotropic (i.e., different energy gaps in different directions), a type of superfluidity that has not been found yet in the laboratory. The energy gap would be of the order of 0.5 MeV (Hoffberg et al., 1970).

4.4. Observational Evidence for Solid State Phenomena in Pulsars

Besides the general slowing down of the pulse rate, two pulsars exhibited a sudden increase in frequency. In March 1969 the Vela pulsar

PSR 0833-45 showed a sudden increase in frequency ("glitch") followed
by the usual (though slightly increased) slowing down (Reichley, Downs,
1969; Radhakrishnan, Manchester, 1969). Two and a half years later
in August 1971 a second glitch of similar magnitude was observed in
Vela (Reichley, Downs, 1971).

In September 1969 a glitch in the Crab pulsar PSR 0531-21 was
observed, smaller but similar in nature to the speed-up inferred for
the Vela pulsar (Boynton et al., 1969; Richards et al., 1969). A second
glitch in the Crab of the same magnitude as the first one was found to
have occurred in October 1971 (Lohsen, 1971).

A detailed analysis of these second glitches is not yet available,
so the following discussion will have to rely on the data evaluated for
the 1969 events.

The parameters for these two pulsars are as follows:

	Vela	Crab
ω	70.5	190
$T = \omega/\dot{\omega}$	2.4×10^4	2.4×10^3
$\Delta\omega/\omega$	2.34×10^{-6}	6.9×10^{-9}
$\Delta\dot{\omega}/\dot{\omega}$	6.8×10^{-3}	8.5×10^{-4}

The analysis of the data showed that the post speed-up behavior
looked very much like some sort of relaxation phenomenon, where the
frequency seems to fall back exponentially to the steady state with a
characteristic time of ~ 8 days for the Crab. Thus the pulsars settled
down to a long-term frequency increase of

$$\Delta\omega/\omega = 0.3 \times 10^{-9} \quad \text{for Crab},$$

$$\Delta\omega/\omega = 2 \times 10^{-6} \quad \text{for Vela}.$$

This post-glitch behavior can be understood in terms of a simple two-
component model (Baym et al., 1969). One component is the combined
crust charged particle system of moment of inertia I_c, rotating uniformly
with the angular velocity $\omega(t)$. The second component is the neutron
superfluid with moment of inertia I_n rotating uniformly with angular
velocity $\Omega_n(t)$. The initial glitch is a sudden change in I_c and $\omega(t)$. Then
the neutron superfluid responds in a characteristic time τ_c to the sudden
change in the crust's angular velocity. If it were just a normal fluid, the
time τ_c, characterizing the coupling between crust and core, would be
of the order of 10^{-10} sec, i.e. no relaxation effects could be observed.
But τ_c is, as stated above, 8 days for the Crab, 1 year for the Vela pulsar,

which very strongly indicates that the interior of these pulsars is super-fluid. When both protons and neutrons are superfluid the coupling is via the magnetic moment interaction of the electrons with the "normal fluid" cores of the vortex lines in the rotating superfluid. This interaction, in fact, has coupling times of the order of 1 year. Thus there is a strong indication that the observation of pulsar glitches allows conclusions on the interior of the neutron star; namely that it contains a crust and a superfluid component.

This is probably a safe conclusion, although there are other theories which try to explain the glitches in a different way:

The Vela glitch is always difficult to account for and most of the theories proposed can be eliminated on the grounds that they offer an explanation only for the Crab pulsar.

The explanation that speed-ups are caused by planets (Rees et al., 1971) is ruled out by the observed post-glitch longterm frequency increase and by the "microglitches" [noise component observed in the Crab pulsar (Groth, 1971)]. Since a critical examination (Börner and Cohen, 1972b) rules out all other explanations, the two-component theory remains the most likely model for the post-glitch behavior.

This model does not account, however, for the sudden initial speed-up and there have been several attempts at explaining that event. The first one proposed was the so-called "starquake-theory" (Ruderman, 1969): The initially oblate crust, formed when the star was spinning comparatively fast and stressed as the centrifugal force on it decreases, cracks when the external stress exceeds the yield point. This results in a fractional decrease in its moment of inertia, and by conservation of its angular momentum in a speed-up of the crust. The magnitude of this effect depends on the magnitude of the stresses that can be built up in the crust. Baym and Pines (1971) used an equation of state developed by Baym et al. (1971b), where nuclei in the lattice become very large, up to $Z \sim 200$. On the other hand, Negele (1973) came to the conclusion that $Z \sim 40$ is to be expected in the crust of neutron stars. The amount of stress that can be built up increases monotonically with Z. It therefore seems that the crustquake theory has serious problems in explaining even the Crab pulsar glitches, if the neutron star crust consists only of small nuclei. Even for $Z \sim 200$ nuclei in the lattice, the requirement to have a typical Crab speed-up every two years leads to a picture of the Crab pulsar as an almost completely solidified star. The crustquake theory predicts a mass of the Crab of less than $0.15\,M_\odot$. This limit should not be taken too much at face value, but it illustrates the difficul-ties of this theory. A neutron star of $0.15\,M_\odot$ might not even be formed in a supernova explosion (cf. Chap. 7), because its binding energy is so low that it becomes energetically more favorable to form dispersed

Fe^{56}. Furthermore $0.15\,M_\odot$ disagrees with all the observational limits discussed in Chap. 7.

In the accretion model (Börner and Cohen, 1971) the initial glitch is ascribed to the infall of material onto the neutron star, transferring angular momentum to the crust and speeding it up. After some time the initial speed-up of the crust is transferred to the interior, and the pulsar settles down to a long-term frequency increase as in the two-component model. Börner and Cohen (1971) find that by choosing a specific model of a neutron star all the unknown quantities are determined, even the infalling mass Δm can be found. Assuming a rotating neutron star model of $1.44\,M_\odot$ they find for the Crab pulsar that $\Delta m = 3 \times 10^{-10}\,m_\odot$, about 10% of the moon's mass. For the Vela pulsar Δm would be $2 \times 10^{-6}\,m_\odot$, about 2/3 of the earth's mass. This theory gives a lower limit from glitch observations of the mass of the Crab pulsar of $1\,M_\odot$.

5. The Liquid Interior

In proceeding to higher densities above $3 \times 10^{14}\,g/cm^3$, we find after the dissolution of the nuclei a mixture of neutrons, protons and electrons, where the protons are just a few percent of the number of neutrons. Although nuclear matter theory is really tested only around normal nuclear density, it is generally believed that the method can be extrapolated with reasonably good results in this regime up to a density of $6 \times 10^{14}\,g/cm^3$ or even $10^{15}\,g/cm^3$.

As the density increases above normal nuclear density, the chemical potentials of the electrons and neutrons increase too. Then when the electron chemical potential (μ_e^-) exceeds the muon (μ^-) rest mass, it becomes energetically more favorable to replace electrons by muons and to start filling up new Fermi levels. The production of π^- mesons would, of course, be even a bigger advantage, since they are bosons, and one could just pack them all into the lowest energy state, thus keeping the electron Fermi energy μ_e^- from ever rising beyond 140 MeV. There are problems, however, because pions and nucleons have a repulsive interaction. In general the S-wave pion-nucleon interaction increases the energy of the π-meson by

$$\delta E = 219\,(\varrho_n - \varrho_p)\,\text{MeV} \tag{5.1}$$

where ϱ_n and ϱ_p are the neutron and proton densities measured in particles per fm^3 (cf. Bethe, 1971). In neutron stars one has mostly neutrons, and $\varrho_n \gg \varrho_p$, and therefore the appearance of π-mesons is postponed to much higher densities.

But in a series of papers Sawyer and Scalapino (Scalapino, 1972; Sawyer, 1972; Scalapino and Sawyer, 1972) have argued that nevertheless a π-meson condensate might form at densities perfectly normal for neutron stars. They have investigated a free Fermi gas of neutrons, and found that the Fermi sea becomes unstable against a π-condensate around a density of $\sim 0.3\,\mathrm{fm}^{-3}$ ($\cong 5 \times 10^{14}\,\mathrm{g/cm^3}$). This leads to a big increase in the proton number density and to a large reduction in pressure, estimated at $\sim 30\%$ in the case of a free Fermi gas by Sawyer and Scalapino. They obtain the result that the π-mesons are formed preferably in a plane wave state of momentum $\sim 170\,\mathrm{MeV}/c$. Large electric currents will flow in the direction of that momentum, and thus they predict a highly directional form of matter – at least over randomly oriented microscopic domains. But the possibility of an appreciable ordering in macroscopic domains cannot be excluded. The effect of this macrostructure will probably be a shifting of the condensate's onset density to higher values.

Sawyer and Scalapino (1972) discuss corrections to their first simple model of a neutron Fermi gas: They include ordinary nuclear forces, emission and absorption of noncondensed pion modes, S-wave pion-nucleon interactions [condition (5.1)], and pion-pion interactions. They came to the conclusion that even condition (5.1) will eventually, when the percentage of protons has increased sufficiently, no longer prohibit the onset of a π-meson condensate. However, they did not carry yet out a detailed quantitative analysis where these corrections are taken into account, and therefore it is unclear at the present time whether the onset of the pion condensate will be shifted to very high densities, or whether it will occur around or below $10^{15}\,\mathrm{g/cm^3}$ with interesting consequences for neutron stars.

Finally also the neutron chemical potential (μ_n) will, with increasing density, become so large that it exceeds the rest masses of the lowest mass hyperons (μ_n here includes the neutron rest mass).

The general formalism describing the appearance of various species of particles i, with number density n, baryon number B_i and charge Q_i, has been given by Ambartsumyan and Saakyan (1960). One has total charge density

$$n_Q = \sum_i n_i Q_i = 0 \tag{5.2}$$

and total baryon density

$$n_B = \sum_i n_i B_i \, .$$

The energy $E(n_1, n_2, ...)$ has then to be minimized with fixed n_B, and $n_Q = 0$. Instead one minimizes

$$E^1 = E - \lambda_n n_B + \lambda_e n_Q. \tag{5.3}$$

If a species i is present E^1 will be a minimum for some $n_i \neq 0$

$$0 = \frac{\partial E^1}{\partial n_i} = \frac{\partial E}{\partial n_i} - \lambda_n B_i + \lambda_e Q_i. \tag{5.4}$$

The Lagrangian multipliers can be determined from the fact that there always are neutrons and electrons present and therefore

$$\lambda_n = \mu_n \quad \text{and} \quad \lambda_e = \mu_e. \tag{5.5}$$

We thus have for the chemical potential of particle i:

$$\mu_i = B_i \mu_n - Q_i \mu_e \tag{5.6}$$

corresponding to the reaction

$$(Q_i) e^- + (\text{particle } i) \leftrightarrow (B_i) (\text{neutron}) + v. \tag{5.7}$$

If a species i is not present, then E^1 always increases as n_i increases, i.e.,

$$\mu_i > B_i \mu_n - Q_i \mu_e. \tag{5.8}$$

The density where the right hand side of (5.8) equals μ_i is the threshold for the appearance of particle i. For free particles μ_i would just be $m_i c^2$; but with interacting particles the threshold density may be lowered in the case of attraction or raised in the case of repulsion. It is generally assumed that the first hyperons make their appearance in neutron star matter at densities around $\varrho_3 \approx 10^{15}$ g/cm^3. The regime of neutron star matter above this density then poses the interesting problem of describing the properties of a system of interacting hyperons and nucleons at zero temperature.

6. The Hyperon Core

6.1. General Remarks

At densities around 8×10^{14} g/cm^3 the forces between the nucleons change from attractive to repulsive, and the precise shape of the repulsive core is important; whereas, for nuclear physics the exact knowledge of the repulsive core does not play an equally important role. The fitting of potentials to reproduce scattering data fixes the attractive region of the potential, but, unfortunately, the shape of the repulsive core is not determined to the same accuracy. In addition to these difficulties, various hyperons appear successively in the neutron matter and their interactions with nucleons and between themselves have to be included in a description of matter at these densities, although they are experimentally very badly known. Furthermore it is not clear up to what densities the nonrelativistic treatment of the interactions in the spirit of nuclear physics is still valid to reasonable accuracy. Some authors (Buchler and Ingber, 1971) believe that already at 10^{15} g/cm^3 non-relativistic many-body calculations break down. It is, of course, extremely important to learn more about the equation of state at these very high densities, since the more massive neutron stars have cores, or consist to a large extent, of very dense matter of roughly 10 times normal nuclear density. Previous estimates (Langer and Rosen, 1970) indicated that the equation of state is changed by less than 1% in the pressure by the inclusion of hyperons, as compared to an equation of state in which their presence is ignored. The basic properties of the nucleon-nucleon interaction at greater than normal nuclear density therefore seem to be more important than the precise statistical equilibrium composition.

6.2. The "Bootstrap" Approach

One of the two main ideas leading to a quantitative description of very dense ("ultradense") matter is, in contrast to what we just said above, based on the speculation that there may exist an "ultradense" region of cold matter, where heavier baryons dominate and where there are so many different types of baryons that only certain statistical features of their distribution and interaction are significant, while the lack of knowledge of the individual interactions is unimportant. It is claimed that by taking into account all the baryon species and their resonances (baryon number $B = 1$ spectrum) and treating them as free particles, a good description of this region is obtained, which also takes into account, by considering all the resonances (the width of resonances is neglected), those features of the interaction important at ultrahigh densities (Frautschi et al., 1971; Wheeler, 1971; Leung and Wang, 1971). The baryon level density is assumed to rise exponentially (Frautschi et al., 1971) as

$$\frac{d \text{ (number of baryon species)}}{d \text{ (mass interval)}} = \varrho_{B=1}(m) \approx m^a e^{bm}$$

$$-7/2 \leqq a \leqq -5/2 \quad b \sim (160 \text{ MeV})^{-1}$$

(6.1)

a form which is suggested by several versions of the "bootstrap" concept of elementary particle theory (Hagedorn, 1968); "Veneziano"-model (c.f. Leung and Wang, 1971).

The equation of state derived from (6.1) becomes extremely soft, the pressure is kept low, even for Fermi-particles, by the effect that with rising density new kinds of particles and resonances are produced with only slightly higher masses, so that only very few Fermi levels are occupied for one type of particle. The equation of state can be derived in analytic form and reads (Wheeler, 1971)

$$P = \varrho c^2/\ln(\varrho/\varrho_0), \qquad (\varrho_0 = 2.5 \times 10^{12} \text{ g/cm}^3).\tag{6.2}$$

The domain of validity of this equation of state does, of course, not necessarily reach down to densities around 10^{15} g/cm^3, where there are just the first few species of known baryons and the statistical formula for the level density is not applicable. Thus Frautschi et al. point out that their formula may only be valid above 10^{17} g/cm^3. This argument then leaves essentially all the problems for neutron star matter unsolved, because stable neutron stars contain matter only up to densities of about 10^{16} g/cm^3 (for the most extreme case of Leung and Wang, 1971). But if one believes that the equation of state behaves asymptotically as predicted by the bootstrap concept, then one could follow Leung and Wang (1971) and apply this concept in the whole region where hyperons are present. The equation of state at lower densities then becomes just the equation of state of a mixture of different noninteracting Fermi gases, until it is joined smoothly at about 5×10^{14} g/cm^3 to the equation of state derived from nuclear many-body theory. In Fig. 2 we plot two equations of state (5 and 6) derived by Leung and Wang (their number I and II; their numerically derived relation is $P \sim \varrho c^2$ at large densities, but the logarithmic dependence in (6.2) might not show up till still higher densities have been reached), where in both cases the bootstrap concept is employed. In (I) a net attractive baryon-baryon force is assumed up to 10^{17} g/cm^3, and the equation of state is joined to a neutron matter equation of state computed from the potential of the Lomon-Feshbach boundary condition model. In (II) the repulsion given by the Reid soft-core potential in neutron matter calculation is assumed to dominate at intermediate densities. Both cases assume complete compensation between repulsive and attractive effects at high baryon densities ($> 10^{17}$ g/cm^3).

The approach discussed above is quite contrary to the spirit of nuclear physics. If the repulsive core in the nucleon-nucleon interaction, for which there is some evidence, persists to very small interparticle separations, then it will become the dominant effect for a wide range of densities, when Fermi statistics and the Pauli principle lose importance because of the many kinds of particle species available. It might be an interesting problem to investigate whether, even in the case of repulsion, the bootstrap concept might take over in a region of extremely high density. I would expect this density to be unrealistically high, however, maybe 10^{30} g/cm^3 or so. This naturally would rob the bootstrap concept of any validity in the case of cold, dense matter.

Although no experimental evidence exists at very high densities, there are difficulties in arguing the repulsive forces away, and if they stay, then the bootstrap approach is neglecting the dominant feature at high densities and is therefore inadequate.

6.3. Manybody Treatment of a Hyperon Gas

The many-body treatment of repulsive baryon-baryon interactions suffers from several uncertainties. The shape of the repulsive core is not known very precisely and no reliable method of computation is available so far. Since it seems to be rather difficult to estimate the effects of higher-order, many-body correlations, which become more and more important with increasing density as nuclear-matter calculations indicate, the consideration of only pair correlations in the approaches so far is questionable too. The use of a potential to describe the interaction between the baryons is certainly better justified here than in high energy particle physics, because one is dealing with matter of high pressure, but low, nonrelativistic kinetic energies.

Both in the approach of Pandharipande (1971) and Bethe and Johnson (1973), which are so far the only attempts to deal with repulsive baryon-baryon interactions using many-body methods, the Reid soft-core potential is used and universal repulsion is assumed in all pairs of interacting particles. Different potentials are assumed in the different angular momentum states $l = 0$, l odd, l even $\neq 0$. Hyperonic matter is assumed to be electrically neutral and the Coulomb force is neglected. The nonrelativistic Schrödinger equation is then employed (the use of the nonrelativistic equation seems to be justified, because it turns out that the momenta of the particles are always small; $v = \frac{1}{2}c$ is the maximum velocity occurring), together with a variational method to numerically minimize the energy, where the trial wave function is a Jastrow-type wave function

$$\chi = A \, e^{ik \cdot r} \prod_{l,m} f_{lm}(r) \tag{6.3}$$

f_{lm} describes the correlation between particles. Only pair correlations are considered. Pandharipande (1973) tested the method by applying it in a computation of the groundstate properties of He^3 and He^4 and obtained excellent results. Although this test speaks in favor of the method, the whole scheme depends very sensitively on the assumptions

Table 3. High density equations of state derived from the Reid soft-core potential. Listed is the energy per particle in MeV as a function of density. BJ: Bethe and Johnson (1972); A, C, N: Pandharipande (1971) Model (A), (C), Neutron matter

Density (g/cm³)	1.7 E 15	4 E 15	1.0 E 16
BJ	250	875	2775
A	22	115	1220
C	100	380	1670
N	157	620	2070

made about the interaction. This is illustrated by Table 3, where the results of models (A) and (C) and of the pure neutron matter calculations of Pandharipande (1971) are compared to Bethe's and Johnson's results.

In his model (A) Pandharipande treats a mixture of the following particles: n, p, Λ, Σ, Δ, μ^-, e^-, assuming universal repulsion and universal intermediate range attraction in all hadron pairs. In model (C) the attraction between hadrons and nucleons is ad hoc lowered by $\sim 10\%$. Bethe and Johnson differ from Pandharipande only in that they assume identical repulsion in S- and P-states, i.e., they take the S-wave repulsion of the Reid soft-core potential ($\sim 6484.2\, e^{-7\mu r}/\mu r$ MeV) for the P-state too, whereas Pandharipande just takes the P-state Reid potential averaged over j (repulsion $\sim 4152.2\, e^{-6\mu r}/\mu r$ MeV). Thus at short distances the repulsion in P-states with Bethe and Johnson is doubled compared to Pandharipande. The differences in the results are quite drastic (c.f. Table 3). Bethe and Johnson obtained energies per particle in excess of Pandharipande's for neutron matter; and, therefore, their equation of state is very stiff and close to the pure neutron matter equation of state. This a rather satisfactory result since it supports the statement made above that for regions where the repulsive forces dominate, nuclear forces determine the equation of state and the nature of the statistical equilibrium composition plays only a secondary role. Bethe and Johnson find an analytic expression for the dependence of pressure and density:

$$P \sim \varrho^{2.54}. \tag{6.4}$$

Pandharipande's model (A) gives negative pressures (for baryon number densities $\leq 1\, \text{fm}^{-3}$) in a certain range of densities, so the possibility of a phase transition accompanying the transition from nuclear to hyperonic matter might be envisioned. In Bethe's and Johnson's approach, as well as in Pandharipande's model (C), however, the transition is smooth. These differences clearly illustrate that one is walking on rather unsafe grounds, and that the precise shape of the repulsive core has a big effect on the equation of state.

6.4. Model of a Lattice of Baryons

It has first been suggested by Bethe (1969) that because of their repulsive interaction at high densities nucleons (and possibly hyperons) would tend to minimize their energy by arranging themselves in some kind of lattice. Similar to the Coulomb lattice of nuclei, which comes into existence because of the strong Coulomb repulsion of the nuclei, this lattice would exist when strong repulsion dominates the interaction. The suggestion is especially attractive with regard to the treatment of higher order correlations in a many-body description of a system of nucleons and hyperons. The lattice structure could be expected to take care of all the higher order terms, and only two-body correlations in a lattice would have to be computed.

First approaches to the problem considered a lattice formed only with neutrons, interacting via the Reid potential (Banerjee et al., 1970); A bcc lattice was then treated as a classical system of oscillators. It was found that the vibrational zero-point energy increased very rapidly with density; and, at 3×10^{15} g/cm^3 the neutrons were already highly relativistic (their vibrational energy exceeded their rest mass). So the classical approach was abandoned. But subsequent quantum-mechanical treatments (Bethe and Johnson, 1973; Pandharipande, 1971) of a lattice of neutrons also ran into difficulties. Even with sophisticated variational techniques the zero-point vibrations became very large. Although the neutrons did not move relativistically here, still their kinetic energies were always at least 40% of the correlation energies that were available to hold them in their lattice positions [at 3×10^{15} g/cm^3 the kinetic energy still exceeds the correlation energy in the calculations of Pandharipande (1971)]. Since it is generally believed that a lattice starts to melt when the kinetic energy of its constituents is 1% of the lattice interaction energy, these results indicate that a lattice will never form at high densities. The nuclear interaction is probably not repulsive enough to force the particles into lattice positions.

There is, however, considerable uncertainty in the microscopic description of matter at these high densities, and many different approximations seem possible. Canuto and Chitre (1973) recently found that a lattice of baryons can exist and will be stable against melting above a density of about 1.5×10^{15} g/cm^3. Their approximation consists of treating only two-particle correlations and cutting off the contributions from all states with angular momentum $J > 2$. Therefore it is possible to have a baryon lattice, if one sticks to a specific approach. The merits of Canuto's and Chitre's work could be judged if the correct description of matter at high densities was known. At the moment, therefore, one can only state that it is an undecided question whether or not a lattice of baryons will exist in the cores of neutron stars.

A microscopic description of a lattice mode of neutrons, protons and other hyperons involves the variation of the number densities of the different particles to find the minimum of the energy. But each time one particle is changed into another species, the symmetry of the lattice will be changed too. In view of these difficulties, Canuto and Chitre (1973) computed only various specific examples with given particle composition and given lattice symmetry. They find that for any given particle composition the fcc lattice always had the lowest energy. They considered as baryon species n, p, Λ, Σ, and found that a lattice consisting purely of lambda particles had the lowest energy. Thus, according to their results, the heavier neutron stars would mainly consist of a core of Λ particles in an fcc lattice, and the name "lambda" star might be appropriate.

Recently it has been claimed (Anderson and Palmer, 1971; Clark and Chao, 1972) that by extrapolating experimental results on quantum solids, one can show that neutron star matter solidifies at densities around 3×10^{14} g/cm^3. This value throws considerable doubt on the validity of the extrapolation. At 3×10^{14} g/cm^3 nucleon-nucleon interaction is still mostly attractive, so there is no reason at that density for an interacting neutron gas to solidify under pressure; since a free Fermi gas would not become solid either. Anderson and Palmer (1971) and Clark and Chao (1972) seem to be unfamiliar with work on neutron lattices discussed above, and it has been shown (Canuto and Chitre, 1973) that their lattice is unstable against melting.

6.5. Conclusion

In Fig. 2 the various equations of state discussed have been plotted. All equations of state which take into account a baryon-baryon potential which is attractive at intermediate range and repulsive at short distances show a qualitatively similar behavior. They are below the line for the free neutron gas up to slightly above normal nuclear density; then they become much stiffer and increase rather sharply in the region of repulsive nuclear forces. Typical for that are curves 2 and 3 which are derived from the Reid potential using nuclear many-body techniques. The equation of state of Bethe and Johnson (1973) is not plotted. It would run slightly above 2 and below 1.

The equation of state numbered 1 (Cohen et al., 1969) crosses the free neutron gas line at a lower density, $\varrho = 2 \times 10^{14}$ g/cm^3, than all the others, which would correspond to a potential changing from attractive to repulsive at interparticle separations greater than 1.5 fm. This equation of state therefore presents something like an upper limit on the pressure vs. density relation; i.e., one cannot expect a many-body calculation with a realistic soft-core potential to lead to a stiffer equation of state. Indeed, the computation of neutron star models from 1 gives, to good accuracy, the same maximum mass models as an approach where constant density throughout the star was assumed, with central pressure and density related through 1.

The equations of state of Leung and Wang (1971), on the other hand, are always below the free neutron gas and have a smaller slope throughout. The departure from all the other equations of state plotted in Fig. 2 becomes very pronounced in the region of normal nuclear density. However, high density matter that soft cannot supply enough pressure to allow the existence of large enough stable neutron stars to agree with astrophysical observations.

The ultimate equation of state will be much stiffer than 5 and 6, less stiff than 1, probably close to 2.

7. Neutron Star Models and Pulsar Observations

7.1. Einstein's Equations

In the preceding sections the nuclear physics of cold dense matter has been discussed, and equations of state of the form $P = P(\varrho)$ (pressure P vs. density ϱ) have been obtained. The procedure leading to neutron star models, however, cannot simply be the integration of the complete set of Einstein's equations – 10 non-linear simultaneous partial differential equations. No one has solved this complete set of equations on the computer. All attemps to do this have led nowhere.

In the introduction we mentioned the more successful approach of Oppenheimer and Volkoff (1939), who used results of Tolman (1939). They reduced Einstein's equations to a simple form for the case of spherical symmetry. Generally denoted by TOV, the equations they used are obtained by eliminating ϕ from the following set

$$\partial_r m = 4\pi r^2 \varrho$$
$$c^2 \partial_r \phi = G(m + 4\pi r^3 p c^{-2}) r^{-1} (r - 2Gmc^{-2})^{-1} \tag{7.1}$$
$$\partial_r p = -(\varrho + pc^{-2}) c^2 \partial_r \phi .$$

We find the above set more useful, however, since it gives the red shift $Z = e^{-\phi} - 1$ automatically. In these equations m is the gravitational mass, ϕ is a relativistic generalization of the Newtonian gravitational potential (it is equal to the Newtonian value in the weak field limit), and r is a radial coordinate chosen in such a way that the surface area of a sphere is $4\pi r^2$. The models obtained are non-rotating and since pulsars are presently believed to be rotating neutron stars, further information is needed.

As will be seen below, typical neutron stars have a radius only a few times their gravitational radius. Thus, general relativistic effects are quite important and must be taken into account. One such general relativistic effect is the dragging along of inertial frames by rotating bodies (Brill and Cohen, 1966; Cohen, 1965). Unlike in Newtonian mechanics, where a gyroscope points towards the same distant star independent of the motion of nearby masses, rotating masses in general relativity drag along the inertial frames and cause the rotation axis of a gyroscope to precess. To include effects like this without having to solve the full set of Einstein's equations a method, valid for slowly rotating stars was developed.

This method (Brill and Cohen, 1966; Cohen, 1965) requires the solution of only one linear equation once Eq. (7.1) are integrated. It is useful for treating pulsars since even the Crab pulsar can be considered a slowly rotating object in the sense that the velocity of any element of the star is small compared to the light velocity and the centrifugal force on the surface is small compared to the gravitational force.

The equation to be solved to treat rotating neutron stars is (Brill and Cohen, 1966; Cohen, 1965; Cohen and Bill, 1968)

$$[A^{-1}B^{-1}r^4\Omega_r]_r = -16\pi BA^{-1}(\varrho + pc^{-2})(\omega - \Omega)Gc^{-2}. \tag{7.2}$$

The quantity Ω is the angular velocity of inertial frames along the rotation axis where it can be measured by observing the precession rate of the axis of a gyroscope. Except for Ω, all quantities are known from Eq. (7.1) or are given. To simplify the equations we have used the quantities A and B defined as

$$A = e^\phi, B^{-2} = 1 - 2Gmr^{-1}c^{-2}.$$

The subscripts r denote differentiation with respect to r and ω is the angular velocity of the star. As boundary conditions on Ω we have $\Omega \sim$ constant near the origin and $\Omega \sim r^{-3}$ outside the star.

Once Ω is determined, it is straightforward to compute quantities of astrophysical interest such as the angular momentum J and the rotational energy E_{rot}. The fully relativistic expression for the moment of inertia of a uniformly rotating body is (Cohen and Cameron, 1971)

$$I = (8\pi/3) \int c\varrho r^4[(1 + p\varrho^{-1}c^{-2})BA^{-1}(1 - \Omega\omega^{-1})] \cdot dr. \tag{7.3}$$

This expression differs from the corresponding Newtonian one by the quantity in brackets. Note that the pressure as well as the density contributes. Also the motion of inertial frames Ω, the red shift ($z = e^{-\phi} - 1$ $= A^{-1} - 1$) and space curvature enter into this general relativistic expression (Cohen and Cameron, 1971). Use will be made of these Eqs. (7.2), (7.3) once non-rotating models have been discussed using Eq. (7.1).

7.2. Properties of Neutron Star Models

Numerical integration of Eqs. (7.1) gives the parameters of neutron star models for various equations of state. Figure 3 shows the variation of the gravitational mass with density. As expected from their very soft equation of state, the neutron star models of Leung and Wang (1971) have very low mass. The maximum mass of their models is not only lower than any of the others shown, but it is also lower than that of Oppenheimer and Volkoff (1939) who neglected repulsion due to strong interactions. If Leung and Wang are excluded from the graph, then agreement between the remaining curves is quite reasonable. Each of these remaining curves CCLR (Cameron, Cohen, Langer, Rosen, 1970), BJ (Bethe, Johnson, 1973) BBS (Bethe, Börner, Sato, 1971) and BPS

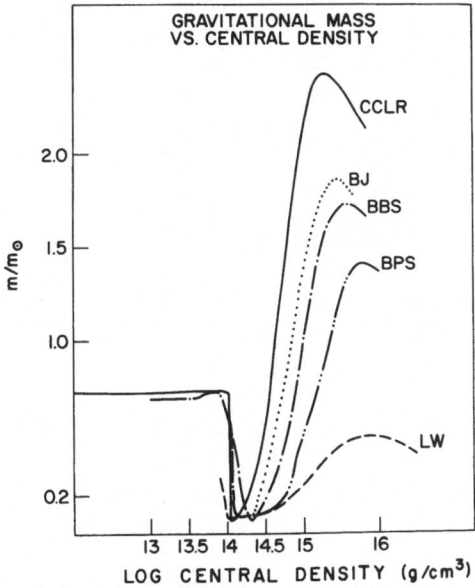

Fig. 3. Gravitational mass vs density for various equations of state

(Baym, Pethick, Sutherland, 1971) gives a maximum mass higher than that of a free Fermi gas. Such behavior is not surprising because of the repulsive core of hadron interactions.

The CCLR equation of state was obtained using the Levinger-Simons velocity dependent potentials while the later calculations BBS, BPS, BJ used the Reid soft core potential. In the high density region above $\sim 2 \times 10^{14}$ g/cm³, use was made of Pandharipande's pure neutron results by BBS and his hyperon results by BPS.

Figure 4 gives the density distribution of selected neutron star models. Note the kink in the curves at $\varrho_k \sim 10^{11}$ g/cm³. This is the point where nuclei become unstable against the emission of neutrons; the neutron drip causes the equation of state to be quite soft in the density region from just above this density ϱ_k to the point where the neutron concentration is sufficient to give a sizeable pressure from the degenerate neutrons. From a plot of the adiabatic index Γ vs. ϱ (Cohen and Cameron, 1971) this effect manifests itself as a rapid decrease in Γ at $\sim 3 \times 10^{11}$ g/cm³.

The angular velocity of inertial frames along the rotation axis as a function of radius is depicted in Fig. 5 for the BBS equation of state. Note that, near the center of the uniformly rotating neutron star, the inertial frames can rotate with angular velocity $\sim 70\%$ that of the star, dropping to $\sim 30\%$ near the surface.

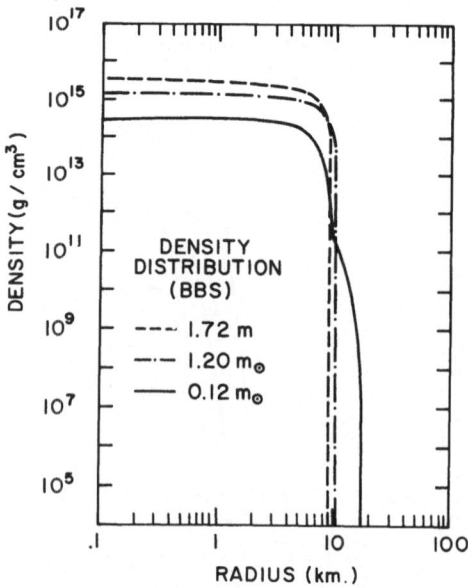

Fig. 4. Variation of density with stellar radius for a selected model based on the BBS equation of state

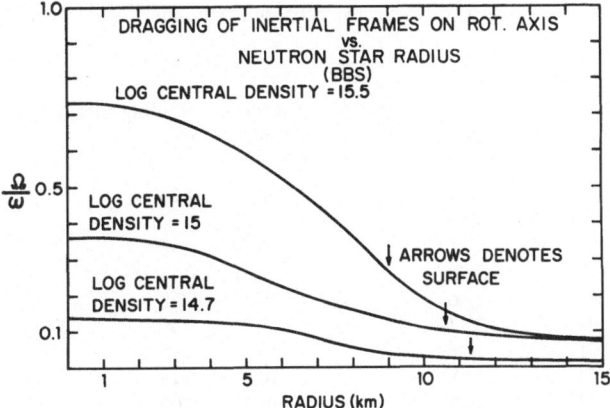

Fig. 5. Angular velocity of inertial frames as a function of radius for models based on the BBS equation of state

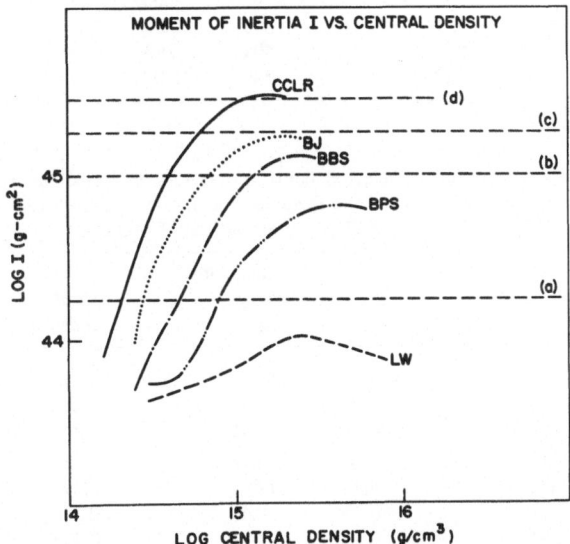

Fig. 6. Moment of inertia (curved lines) of neutron star models as a function of density for various equations of state

Such large dragging of inertial frames makes a significant contribution to the moment of inertia as can be seen from Eq. (7.3). The moment of inertia is plotted in Fig. 6 for various equations of state. An interesting property of the CCLR, BJ, and BBS curves is that the moment of inertia peaks at densities below the mass peak. Hence the addition of material

Table 4. Equilibrium properties of selected neutron star models based on BBS equation of state

log central density g/cm³	Mass/10³³ g	Proper mass/10³³ g	Radius km	Binding energy/10³³ g	Percent binding
15.6	3.45	4.69	8.74	1.24	26.4
15.5	3.44	4.56	9.13	1.12	24.5
15.4	3.35	4.31	9.54	0.96	22.2
15.3	3.17	3.95	9.92	0.78	19.7
15.2	2.88	3.46	10.2	0.58	16.9
15.1	2.48	2.87	10.5	0.40	13.9
15.0	2.01	2.25	10.6	0.25	11.0
14.9	1.53	1.68	10.7	0.14	8.4
14.8	1.09	1.16	10.7	0.07	6.1
14.7	0.73	0.77	11.0	0.03	4.2
14.6	0.52	0.54	11.6	0.02	3.0
14.5	0.37	0.38	12.5	0.01	2.2
14.4	0.25	0.25	15.1	0.004	1.6
14.3	0.14	0.14	47.9	0.001	0.8
14.2	0.37	0.45	341	0.03	6.5
14.1	0.95	1.03	401	0.03	2.7
14.0	1.10	1.17	349	0.02	1.9

Table 5. Pulsational properties of selected neutron star models based on BBS equation of state. T_0 is the pulsation period of the fundamental mode. The pulsational energy is given in units of $10^{52}\xi^2$, where $\xi = \delta R/R$ is the amplitude of pulsation

log central density g/cm³	Mass/10³³ g	T_0 msec	Pulsational energy/10⁵²ξ² ergs	Central redshift	Surface redshift
15.6	3.45	U	U	2.31	0.551
15.5	3.44	0.79	1.41	1.84	0.506
15.4	3.35	0.51	4.91	1.42	0.446
15.3	3.17	0.41	8.78	1.08	0.379
15.2	2.88	0.35	12.8	0.797	0.309
15.1	2.48	0.33	18.4	0.575	0.241
15	2.01	0.31	13.9	0.405	0.179
14.9	1.53	0.31	15.6	0.282	0.127
14.8	1.09	0.33	16.1	0.190	0.085
14.7	0.73	0.38	3.01	0.124	0.053
14.6	0.52	0.45	18.6	0.088	0.035
14.5	0.37	0.68	237	0.064	0.023
14.4	0.25	1.40	935	0.044	0.012
14.3	0.14	168.0	3.47	0.025	0.0021
14.2	0.37	U	U	0.021	0.0011
14.1	0.95	U	U	0.020	0.0019
14	1.10	U	U	0.019	0.0022

to massive (stable) neutron stars can reduce the moment of inertia – a property which does not depend on a particular equation of state.

The horizontal lines a and b represent lower limits on the moment of inertia which can be obtained from comparison with observation (cf. 7.3).

Table 4 (from Börner and Cohen, 1973) gives the equilibrium properties of a number of models for the BBS equation of state. A remarkable feature of all these models is their large binding energy – reaching values as high as 25% of the rest mass energy.

The pulsational properties of neutron stars are given in Table 5 (Börner and Cohen, 1973), together with the redshifts. The pulsational energy is obtained by multiplying the values given in the table by $10^{52} \xi^2$. Here ξ is the relative change of radius $\delta R/R$.

For central densities less than 3×10^{14} g/cm^3 the models have very high pulsational energies. Even $\xi = 0.1$ oscillations have energies of 2×10^{52} ergs to 10^{53} ergs, in excess of the binding energy of these neutron stars. Thus these stars when formed will probably convert all the energy available into oscillations and not evolve into a rotating object. In the fundamental mode a typical neutron star pulsation period is ~ 0.4 msec.

Table 6. Rotational properties of selected neutron star models based on BBS equation of state

log central density g/cm^3	Mass /10^{33} g	Crust mass /10^{31} g	Moment of inertia /10^{45} g cm^2	Moment of inertia of crust /10^{43} g cm^2	Central dragging of inertial frames Ω_C/ω	Surface dragging of inertial frames Ω_S/ω
15.6	3.45	0.790	1.19	0.52	0.79	0.26
15.5	3.44	0.995	1.26	0.70	0.73	0.25
15.4	3.35	1.30	1.30	0.96	0.68	0.22
15.3	3.17	1.72	1.27	1.31	0.61	0.19
15.2	2.88	2.31	1.16	1.78	0.53	0.16
15.1	2.48	3.11	0.98	2.33	0.45	0.13
15	2.01	4.12	0.75	2.91	0.36	0.093
14.9	1.54	5.31	0.53	3.40	0.28	0.064
14.8	1.09	6.70	0.33	3.74	0.21	0.040
14.7	0.73	8.35	0.20	4.00	0.14	0.022
14.6	0.52	9.56	0.13	3.94	0.11	0.012
14.5	0.37	10.4	0.08	3.67	0.08	6.5 D − 3
14.4	0.25	10.8	0.05	3.07	0.06	2.1 D − 3
14.3	0.14	11.5	0.04	3.87	0.03	5.4 D − 3
14.2	0.37	36.5	76.2	76.2	0.03	3.9 D − 4
14.1	0.95	94.7	378	378	0.03	8.7 D − 4
14	1.10	110	396	396	0.02	9.6 D − 4

Table 6 (Börner and Cohen, 1973) gives the crust's moment of inertia and that of the entire star vs. central density. The star's crust (where a nuclear lattice exists) is assumed to reach to a density of about 2×10^{14} g/cm^3 (Baym et al., 1971 b).

7.3. The Crab Pulsar

The Crab pulsar is located at the center of the Crab nebula, the remnant of a supernova that exploded in 1054. Because of its location in the nebula a distance estimate independent of the pulsar's dispersion can be obtained. But Trimble and Woltjer (1971) have pointed out that the distance of the Crab is still quite uncertain and that values as low as 1.2 kpc or as high as 2.5 kpc cannot be excluded.

Let us, however, be definite and adopt a distance of 2 kpc for the Crab Nebula. This value seems to be the most widely used. Observations of the nebula indicate that energy must be supplied to it continuously. Assuming that the Crab pulsar is the only source of energy in the nebula, one can determine limits on the energy output of the pulsar (Rees and Trimble, 1970; Börner and Cohen, 1972) by considering the energy balance.

The only well established energy loss is the synchrotron radiation which implies $\dot{E}_{synch} = 1.2 \times 10^{38}$ erg/sec (Baldwin, 1971), if the distance to the Crab is 2 kpc. The pulsar has to replenish at least the electrons producing the optical and X-ray synchrotron radiation, because these particles have half-lives of less than 100 years. So a rough estimate obtained from the observed spectrum (Baldwin, 1971) indicates that the pulsar has to supply continuously at least 0.8×10^{38} erg/sec. For the pulsar to replenish this energy, via loss of rotational energy

$$\dot{E}_{rot} = I\omega\dot{\omega} ,$$

its moment of inertia has to be at least 1.8×10^{44} g cm^2 (c.f. Cohen and Cameron, 1971), corresponding to line (a) in Fig. 6. Neutron star models corresponding to various equations of state have been discussed above and it can be seen from Fig. 3 that a model with $I = 1.8 \times 10^{44}$ g cm^2 has a mass of

$$0.34\, M_\odot\,(BPS), \quad 0.36\, M_\odot\,(BJ), \quad 0.36\, M_\odot\,(BBS), \quad 0.26\, M_\odot\,(CCLR).$$

Figure 6 also shows that the equation of state of Leung and Wang (1971) cannot provide neutron star models big enough to exceed this lower limit.

The evidence in favour of massive neutron stars becomes even stronger if protons are taken into account. If protons that are pulled from the surface of the rotating neutron star (Goldreich and Julian, 1969) are accelerated to the same energies as the electrons producing synchrotron radiation the minimum energy loss is twice the synchrotron radiation loss. Then the minimum moment of inertia of the Crab pulsar increases by a factor of 2:

$$I_{min} \geqq 4 \times 10^{44} \text{ g cm}^2 .$$

The neutron star mass is then $\geqq 0.5 \, M_\odot$ (BBS, BJ).

The acceleration mechanism of Gunn and Ostriker (1969) can also be tested. They require that the protons get ten times the energy of the electrons. Assuming the fluxes of electrons and ions from the pulsar to be equal, this would require ten times the synchrotron energy for the protons, leading to a moment of inertia of 2×10^{45} g cm^2 (line (c) in Fig. 6). Only the CCLR equation of state has models with moments of inertia of that magnitude. But even then this condition is satisfied only over a small density range near the mass peak. It therefore seems that the model of Gunn and Ostriker (1969) should be modified quantitatively.

Conclusions based on energy losses from the Crab nebula due to the expanding supernova shell (Rees and Trimble, 1970; Börner and Cohen, 1972) are much more uncertain than the preceding considerations. Observations of the filaments in the expanding supernova shell (Woltjer, 1958; Trimble, 1968) show that the expansion velocity at present is higher than would correspond to an expansion at constant velocity since 1054. It seems that the nebula is accelerating now with an acceleration of

$$\dot{v} = 0.0014 \text{ cm/sec}^2 .$$

The nebula might, however, be decelerating now with the velocity still higher than the average, if it had been accelerating rather strongly in the past. The energy of the expanding supernova shell would change due to this acceleration at a rate

$$\dot{E}_{acc} = \int \varrho v \dot{v} \, d(\text{Vol}) \, (\varrho = \text{density in the nebula}) . \tag{7.4}$$

It would also change by the "snow plow" effect, the change in mass of the supernova shell as interstellar material piles up along the rim

$$\dot{E}_{plow} = \tfrac{1}{2} \varrho_m v^3 A \quad (A = \text{surface area of nebula}, \varrho_m = \text{density of interstellar}$$
$$\text{material}) . \tag{7.5}$$

Since we do not know whether the supernova shell is accelerating or decelerating at present, we investigate both cases. If the shell is decelerating, the energy gained by deceleration will be spent in the "snow plow" effect described by (7.5), and perhaps totally balance it. Thus $\dot{E}_{shell}$ $= \dot{E}_{acc} + \dot{E}_{plow} = 0$ is a distinct possibility. No further limits on the parameters of the Crab pulsar except those derived earlier from synchrotron radiation can be found in this case. We should notice, however, that in principle one could directly measure the value of $\dot{v}$ at present, and thus decide the question of acceleration or deceleration. If the currently accepted values for acceleration and snow-plow are used, we find

$$\dot{E}_{acc} = 1.6 \times 10^{38} \text{ erg/sec}, \tag{7.6}$$

$$\dot{E}_{plow} = 1.7 \times 10^{38} \text{ erg/sec}, \tag{7.7}$$

hence

$$\dot{E}_{shell} = 3.3 \times 10^{38} \text{ erg/sec}. \tag{7.8}$$

This energy has to be supplied either directly by the pulsar via the low frequency waves emitted or by the adiabatic expansion of a relativistic gas (Trimble and Rees, 1970). In both cases a rotating neutron star with a moment of inertia of 10^{45} g cm^2 can continuously supply that energy [line (b) in Fig. 6]. The equations of state that can furnish a neutron star model with a moment of inertia big enough (BBS, BJ, CCLR) give a mass of $1.2 M_{\odot}$ (BBS, BJ) for this model. If little material was lost during the collapse, the star would have had a mass of $1.35 M_{\odot}$ prior to the collapse, above the Chandrasekhar limit for typical white dwarfs. It should be remembered that these energy losses are rather uncertain, and that the pulsar has to supply the energy continuously only if the energy content of the gas of relativistic particles in the nebula is maintained at its present level. All the uncertainties can, however, be decided by future observations, and thus a value of $1.2 M_{\odot}$ for the Crab pulsar may be confirmed some day with a much higher degree of confidence than we have now.

If the shell is pushed out by low-frequency waves from the pulsar, and if according to Gunn and Ostriker (1969) protons get ten times the energy of the electrons, then only the maximum mass (near the mass peak) neutron star models of the CCLR equation of state can fulfill this requirement, as indicated by line (d) in Fig. 6. This particle acceleration mechanism therefore seems to be unrealistic.

If the distance to the Crab were less or more than 2 kpc, the limits derived above would have to be scaled down or up accordingly. If we go to the extreme values for the distance of 1.2 kpc to 2.5 kpc, the moment of inertia necessary to account for the short-lived synchrotron electrons would vary between 0.8 and 2.4×10^{44} g cm^2 [$0.2 M_{\odot}$ and $0.4 M_{\odot}$, respectively (BBS)].

7.4. The Vela Pulsar

The Vela pulsar PSR 0833–45 is associated with the supernova remnant Vela X, which is about 1.1×10^4 years old, has a radius of 10 pc, and is at a distance of 500 pc (Milne, 1970).

If the electromagnetic radiation emitted by Vela X is synchrotron, then from synchrotron theory the energy content in the gas can be estimated at $\sim 10^{49}$ (Tucker, 1971). Assuming constant velocity of expansion of the supernova shell ($v = 880$ km/sec), the energy loss through adiabatic expansion would be

$$\dot{E}_{ad} = 2.4 \times 10^{37} \text{ erg/sec} . \tag{7.9}$$

On the other hand the rotational energy lost by a neutron star with the parameters of PSR 0833–45 is between

$$\dot{E}_{rot} = 4 \times 10^{36} \text{ erg/sec} \quad \text{to} \quad 8.5 \times 10^{36} \text{ erg/sec} \tag{7.10}$$

for models with mass between $0.8\, M_\odot$ and $1.7\, M_\odot$ (BBS). Although this is rather large compared to the loss of 4×10^{35} erg/sec in X-rays (Tucker, 1971) and 10^{33} erg/sec in radio (Milne, 1970), the pulsar cannot supply the energy given in Eq. (7.9). The shell must therefore have been decelerating. Let us assume that the deceleration at present is very small, and that the main contribution to $\dot{E}_{shell}$ is by "snow plow" (Börner and Cohen, 1972). If it is further assumed that a neutron star of $1.2\, M_\odot$ is present to balance the expansion losses, then the velocity of expansion can be determined to be 240 km/sec (Börner and Cohen, 1972). It is amusing to note that at about the same time Wallerstein and Silk (1971) independently (neither group knew of the other's work until after publication) measured the expansion velocity of Vela X by an observation of CaII lines in that direction, and they found 240 km/sec for the expansion velocity.

The "glitches" of the Vela pulsar ($\Delta\omega/\omega \sim 10^{-6}$) are ascribed to corequakes by Pines et al. (1972), sudden relaxations of stress stored in the neutron star's central core made of a hadron lattice. It is still an open question of whether or not a solid hadron core may exist in neutron stars. The work of Canuto and Chitre (1972) suggests that a solid core might form at densities above 1.5×10^{15} g/cm^3. The corequake theory therefore requires the Vela pulsar to have a central density above this value, which makes Vela a rather heavy neutron star with mass greater than $0.8\, M_\odot$ (BPS), $1.2\, M_\odot$ (BBS; CCLR) and $1.5\, M_\odot$ (BJ), according to the various equations of state. The stress in the core is not built up between glitches in this model, but rather each glitch takes out a small part of a huge reservoir of elastic stress in the core. One has to explain why this produces glitches of $\Delta\omega/\omega \sim 10^{-6}$ instead of a continuous relaxation or one extremely big jump.

8. Neutron Stars Accreting Matter

During the last 2 years galactic X-ray sources showing regular pulsation periods have been discovered by the Uhuru satellite (see 8.6). It is known that these sources are members of close binary systems. The regular pulsations suggest that a clock mechanism is at work, and the length of the periods (1.24 sec for Her–X 1, cf. 8.6) indicates that the rotation of a neutron star provides the time keeping mechanism. The fact that these sources are in binary systems strongly supports earlier suggestions (cf. e.g. Shklovsky, 1967; Prendergast and Burbidge, 1968) that X-ray sources are an accretion phenomenon. The companion star in the binary system provides the mass flux, and the matter releases up to 25% of its rest mass energy by falling to the surface of a neutron star (cf. Chap. 7). Most of the questions concerning accretion on compact objects have not been adequately explored yet, and theories are of a very qualitative nature. In the present chapter, nevertheless, an attempt is made to sort out the different problems, and to present the solutions of some of them. There will be no discussion of complicated phenomena as the gas flow in a binary system; instead the attention will be confined to the immediate surroundings of the neutron star undergoing accretion. We shall discuss single neutron stars accreting interstellar gas, and neutron stars in binary systems. We shall speculate about the influence of strong magnetic fields of the neutron star. The subsequent discussion will be focussed on those aspects of the problem connected with the structure of the neutron star itself – the change of its surface composition by accretion, the nuclear reactions induced by the accreted particles, the X-ray and Gamma ray spectrum emitted, and its possible diagnostic value for directly measuring the physical parameters of neutron stars.

8.1. The Eddington Limit

Consider a single proton falling onto a neutron star. After it has been added to the star, which settles into a new equilibrium configuration with slightly higher mass, and a slightly smaller radius, the total amount of energy released during infall is the difference in gravitational energy between the initial and final configuration. From the neutron star models of Chapter 7, we find that the gravitational binding energy may amount to 25% of the proper mass, for the most massive models. For a typical star of 1 $M_\odot$ the binding energy is 10%. Thus in the process of a proton falling onto the surface of a neutron star of 1 $M_\odot$, the energy of 0.1 $m_p c^2$ ≈ 100 MeV is released. If we now picture a stream of particles with mass flux $\dot{M}$, which release 100 MeV/proton suddenly when they hit the neutron star surface, a temperature corresponding to that mean energy, namely of $10^{12}\,°K$ is created. On the other hand, assuming that the neutron star radiates as a black-body, one finds from

$$L = 0.1\,c^2\dot{M} = 4\pi\,R^2\,\sigma\,T^4 \tag{8.1}$$

a temperature

$$T = \left(\frac{0.1\,c^2\dot{M}}{4\pi R^2 \sigma}\right)^{1/4} = 10^7 \left(\frac{\dot{M}}{10^{17}}\right)^{1/4} \quad \text{with} \quad \dot{M} \text{ in g/sec} \tag{8.2}$$

This discrepancy between these two values is disturbing, and indicates the formidable problem that has to be solved.

From quite general considerations a few limitations on the accretion process on neutron stars can be derived. The luminosity that can be gained from the accretion process is determined by the inflow of mass $\dot{M}$, but it is limited by the fact that the outgoing photons interact with the infalling matter. Eddington (1926) derived a limit for the case of stellar equilibrium and spherical symmetry, and it has been rediscussed for the case of accretion by Zel'dovich and Novikov (1964): The photons scatter mainly on the electrons in the incoming material, and on the average in the scattering process a photon will transfer all of its momentum $h\nu/c$ to the electron. Thus for a spherically symmetric situation on the average an electron at radius r experiences a photon force of

$$K = \sigma_{tot}(L/4\pi r^2 c) \tag{8.3}$$

where σ_{tot} is the total cross section $0.66 \times 10^{-24}\,\text{cm}^2$ if the photon energies in the electron rest frame are small compared to $m_e c^2$; L is the total luminosity of the neutron star. Since σ_{tot} is proportional to m^{-2} protons are not directly affected by this force, but the charge separation produced by the electrons moving slightly outwards relative to the protons creates an electric field. This field transfers the radiation force acting on the electrons to the protons. Thus each infalling proton experiences the drag of gravity and the outward pushing photon force. The net force on each proton (the infalling material will be mainly a gas of ionized hydrogen) is

$$K_p = \sigma_{tot}L/4\pi r^2 c - GMm_p/r^2 \,. \tag{8.4}$$

Here m_p is the proton mass, M the neutron star mass. K_p is zero for a critical luminosity

$$L_c = 4\pi GMm_p c/\sigma_{tot} = 1.3 \times 10^{38} \text{ ergs/sec}\,, \tag{8.5}$$

where we assumed $M = 1 m_\odot$. From (8.1) it follows that to L_c corresponds a mass accretion rate of

$$\dot{M}_c = 1.4 \times 10^{18} \text{ g/sec} .$$

L_c is an upper limit on the luminosity for a stationary and spherically symmetric mass flow, because for L exceeding L_c from (8.4) K_p becomes positive, and the infalling matter will be stopped. But then the luminosity will drop too, and finally this self-regulating mechanism will restrict the average accretion rate to $\dot{M}_c$ (Zel'dovich, 1964). It is an open (and interesting) question whether supercritical accretion $\dot{M} > \dot{M}_c$ could lead to flare phenomena. It seems plausible that the Eddington limit can be exceeded in cases of nonstationary accretion, where a massive object hits the neutron star accreting as optically thick material within the free fall time. Also a nonspherically symmetric accretion of matter may substantially change the upper limit of the accretion rate. In most cases, however, realistic accretion rates will be below $\dot{M}_c$.

8.2. Accretion Rates

Let us, as a first instructive example, consider the accretion of interstellar gas by a single neutron star. Effects of a magnetic field will also be neglected. This problem has been treated by various authors and is discussed in detail in Chapter 13 of the book by Zel'dovich and Novikov (1971): Consider a system of noninteracting particles (with density ϱ_∞, speed v_∞) far from a star. If v_∞ is small compared to the free fall velocity $v = (2GM/R)^{1/2}$, the mass flux onto the star is determined by the impact parameter l_{max}; particles with l_{max} would hit the star's surface tangentially. Thus

$$\dot{M} = \pi r_g^2 c \varrho_\infty (c/v_\infty) (R/r_g) , \tag{8.6}$$

where r_g is the gravitational radius

$$r_g = 2GM/c^2 .$$

Formula (8.6) is valid in the nonrelativistic case, precisely for $R/r_g > 4$. If $R/r_g < 4$, which is the case for massive neutron stars the general relativistic formula must be used, where

$$\dot{M} = 4\pi r_g^2 c \varrho_\infty (c/v_\infty) . \tag{8.7}$$

Or, in a more convenient form

$$\dot{M} = 500 (\varrho_\infty/10^{-24} \text{ g cm}^{-3}) (M/M_\odot)^2 (v_\infty/10 \text{ km sec}^{-1})^{-1} .$$

Since typical densities and velocities of interstellar matter in "HII" regions (regions of ionized hydrogen) are $\varrho_\infty = 10^{-24}$ g cm^{-3}, $v_\infty = 10$ km sec^{-1} this accretion rate is almost negligible. However, the actual flux will be much different, because the interstellar particles interact with each other, and these interactions will reduce the tangential velocities and permit only radial velocities to grow during infall. It has been shown by Shvartsman (cf. Zel'dovich and Novikov, 1971, Chapter 13) that it is valid to describe the infalling particles as a hydrodynamical fluid. A simple justification can be given by taking into consideration the existence of a magnetic field frozen into the infalling plasma. The Larmor radius for a proton moving with thermal velocity is smaller than the characteristic length scale of the infalling matter as long as $B > 10^9\, r^{-3/2}$ gauss; that is a frozen-in interstellar magnetic field $B = 10^{-6}$ gauss will keep the protons linked to the field lines for $r > 10^{10}$ cm. This is much smaller than the typical accretion radius of $\approx 10^{14}$ cm, and hence the hydrodynamical approximation seems to be justified. This means that the interaction between particles has a very large cross section, and it turns out that (see also Thorne and Novikov, 1973)

$$\dot{M} \cong r_g^2 c\, \varrho_\infty (c/v_\infty)^3 , \tag{8.8}$$

i.e. M is higher by a factor $(c/v_\infty)^2 \approx 10^9$ over the case of the inflow of noninteracting particles. In different units (Thorne and Novikov, 1973)

$$\dot{M} = (10^{11} \text{ g/sec}) (M/M_\odot)^2\, (\varrho_\infty/10^{-24} \text{ g cm}^{-3}) (T_\infty/10^4 \text{ K})^{-3/2} . \tag{8.9}$$

Hence a typical accretion rate on a $1 M_\odot$ neutron star in the interstellar medium is $M = 10^{11}$ g/sec. For neutron stars in a binary system the accretion rate can be much higher, and may approach the critical value $\dot{M}_c$.

8.3. The Death of Pulsars

While they are actively radiating as pulsars, single neutron stars in the interstellar gas will not accrete matter. But as they slow down they reach a velocity of rotation below which the interstellar gas can no longer be kept away. The accretion process may extinguish the pulsar and it may explain the absence of pulsars with periods longer than a few seconds. The problem has been considered independently by Meyer (1972) and Shvartsman (1970). We follow here the discussion given by Meyer (1972):

The radiation pressure of a rotating neutron star (ω, R) decreases as r^{-2}, where r is the distance from the star

$$P_R = \overline{B_n^2} R^6 \omega^4 /(4\pi c^4\, r^2) . \tag{8.10}$$

Here P_R is the radiation pressure in the equatorial plane of the axis of rotation; a magnetic dipol field was assumed whose average component normal to the rotation axis is denoted by $\bar{B}_n$.

As P_R decreases with ω^4 the equilibrium with the pressure of the interstellar gas is maintained by the equilibrium surface moving toward the pulsar. But eventually a point will be reached, where the gas is appreciably affected by the gravitational field of the star. Then the gas pressure will be

$$P_G = P_\infty \exp((GM/R_g T)\, r^{-1}) \tag{8.11}$$

(R_g = gas constant, P = pressure at infinity).
Equilibrium $P_G = P_R$ requires

$$\left(\frac{R_g T}{GM}\, r\right)^2 \exp\left(\frac{GM}{R_g T}\, r^{-1}\right) = \frac{\bar{B}_n^2 R^6 \omega^4}{4\pi c^2 P_\infty}\left(\frac{R_g T}{GM}\right)^2. \tag{8.12}$$

The left hand side of (8.12) has its minimal value for

$$r = r_c = GM/(2R_g T). \tag{8.13}$$

If the right hand side is smaller than this minimum value, equilibrium can no longer be achieved. That is, for $\omega < \omega_c$ with

$$\omega_c = \left(\pi(\exp 1)^2\, c^4 (P_\infty/(\bar{B}_n^2 R^6))\, (GM/R_g T)^2\right)^{1/4} \tag{8.14}$$

the pulsar can no longer find an equilibrium surface, and interstellar matter will accrete on the neutron star. The critical period is

$$P_c \equiv 2\pi/\omega_c = 0.9 \sec \left(\frac{P_\infty}{10^{-12.5}\,\text{dyn cm}^{-2}}\right)^{-1/4}\left(\frac{\bar{B}_n}{10^{12}\,\text{gauss}}\right)^{1/2}$$
$$\cdot \left(\frac{R}{10^6\,\text{cm}}\right)^{3/2} (M/M_\odot)^{-1/2}\, (T/100\,\text{K})^{1/2}. \tag{8.15}$$

P_c is of the order of 1 sec, which fits nicely with the observed facts. Thus accretion of interstellar gas may be responsible for the termination of pulsar activity.

The fate of neutron stars in binary systems has been discussed repeatedly, especially since the discovery of periodically pulsating X-ray sources (cf. 8.6). Shvartsman (1971) has pointed out that the stellar wind from the companion star in a binary system can very effectively suppress the ejection of particles from the neutron star, and thus "neutron stars in binary systems should not be pulsars". Pringle and Rees (1972) consider the accretion process (in a binary system) in cases when the infalling matter possesses angular momentum and forms a disc spinning around the neutron star. They point out that accretion can only take place, if the rotation period of the neutron star is above a certain limit.

$$P_{min} = 0.42 \, B_{12}^{6/7} \, (\dot{M_c}/10^{18}) \, (M/M_\odot)^{-5/7} \, R_6^{15/7} \, (B_\phi/B_p)^{3/7} \, x^{3/7} \text{ sec} . \qquad (8.16)$$

Here (B_ϕ, B_p) are the toroidal, resp. poloidal magnetic fields in the disc, x is a parameter of order unity describing the disc's thickness, B_{12} is the magnetic field of the star in units of 10^{12} gauss; R_6 is the star's radius in units of 10^6 cm.

For periods less than P_{min} matter is continuously ejected from the inner edge of the disc by the star's rotating magnetic field. In that case the rotation of the neutron star would, just as it does for pulsars, supply energy to the system. When the period exceeds P_{min} accretion will take place, and provide the energy for the system. Pringle and Rees (1972) show that accretion from a disc will in general speed up the stellar rotation. Similar considerations have been put forward by Lamb et al. (1972), and Davidson and Ostriker (1972).

8.4. The Accretion Process in Detail

While the physics of an accretion disc surrounding a collapsed object (white dwarf, neutron star, black hole) has been discussed by a number of authors (Prendergast and Burbidge, 1968; Cameron and Mok, 1967; Melrose and Cameron, 1968; Pringle and Rees, 1972), the phenomena in the atmosphere and the surface of a neutron star have been investigated only in a preliminary way. The work of Zel'dovich and Shakura (1969) is still the most thorough, to my knowledge, but it suffers from the fact that effects of the star's magnetic field are not treated. The discovery of the pulsating X-ray sources indicates, indeed, that these effects are important, and that the assumption of spherically symmetric accretion may not be realistic. Recently Ramaty et al. (1973), and Börner et al. (1973) have been concerned with spallation reactions induced on the neutron star surface by accreting particles; they also investigated the change in the surface composition by accretion (cf. 8.5 and 8.7).

In this section the paper by Zel'dovich and Shakura (1969) will be presented, although their results may not be directly applicable to the observations, but their work contains interesting points of principle. They consider spherically symmetric accretion of protons on a neutron star of mass $M \approx 1 \, M_\odot$ and radius $R \approx 10^6$ cm. For mass fluxes of

$\dot{M}_1 = 10^{17}$ g/sec, or $\dot{M}_2 = 10^{16}$ g/sec the density ϱ in the impinging stream

$$\dot{M} = 4\pi\varrho v R^2 \tag{8.17}$$

is $\varrho_1 = 5 \cdot 10^{-7}$ g/cm^3, or $\varrho_2 = 5 \cdot 10^{-8}$ g/cm^3 respectively.

The essential point is that the protons in the incident stream are not decelerated instantaneously (corresponding to a temperature of 10^{12} °K), but deposit their energy gradually in an extended layer of the atmosphere and surface. If y denotes the amount of material above a certain depth x

$$y = \int_x^\infty \varrho(x)\,dx,$$

and if y_0 denotes the mean free path to the incident particles, then in a first, linear approximation the energy released per gram of material is

$W = Q/y_0$ for $y < y_0$ and

$W = 0$ for $y > y_\circ$, where $Q = L/4\pi R^2$

is the energy flux per unit surface area of the star. The electrons in the y_0 layer receive energy from the incoming protons, and release energy by bremsstrahlung or through the inverse Compton effect. Both these processes have to be treated as occurring in a nonequilibrium radiation field, where the photon temperature T_v is less than the electron temperature T_e. Zel'dovich and Shakura deal with that problem in a first approximation by introducing effective radiation temperatures for bremsstrahlung T', and inverse Compton effect T'', which are assumed to be equal to T_v. The energy balance requires

$$Q/y_0 = 5 \cdot 10^{20}\, T\varrho(1 - T'/T) + 6.5\,\varepsilon T(1 - T''/T). \tag{8.18}$$

Here T is the electron temperature, ε is the energy density of the radiation, and $T' = T'' = T = (\varepsilon/a)^{1/4}$, where $a = 7.8 \cdot 10^{-15}$ erg/cm^3 deg^4. ε is determined by the requirement, that

$$Q(y - y_0) = -(c/3\sigma)\,d\varepsilon/dy \quad y < y_0, \tag{8.19}$$

where $\sigma = 0.38$ cm^2/g is the opacity of a fully ionized plasma. Equation (8.19) is solved with the boundary condition $\varepsilon = (\sqrt{3}/c)\,Q$ for $y = 0$, and then, by also employing the equation of hydrostatic equilibrium, the density and temperature distribution in this model can be found as a function of the accretion rate (assuming $L = \dot{M}\,0.1\,c^2$). In the interior, where $y \gg y_0$, complete thermodynamic equilibrium is established, and $T = (\varepsilon/a)^{1/4} \sim L^{1/4} y_0^{1/4}$. The equilibrium temperature thus does not depend strongly on y_0, but the temperature and density distribution in the y_0 layer will depend decisively on the mean free path y_0 of the incident particles. Zel'dovich and Shakura consider two distinct possibilities: Firstly, in the case where the incident protons are decelerated by collisions the mean free path for 100 MeV protons is $50 - 100 \, \text{g/cm}^2$. Zel'dovich and Shakura adopt $y_0 = 20 \, \text{g/cm}^2$, a rather low value in my opinion. In the interior $T \sim y_0^{1/4}$, and different values for y_0 do not affect the equilibrium temperature very much. It is found that $T_1 = 1.5 \cdot 10^7 \, ^\circ\text{K}$, resp. $T_2 = 8.6 \cdot 10^6 \, ^\circ\text{K}$ for $y \gg y_0$, and $\dot{M}_1$, resp. $\dot{M}_2$. The electron temperature at the surface, where bremsstrahlung is negligible (because it is $\sim \varrho$) is determined by the inverse Compton effect as $T_e \approx 10^8 \, ^\circ\text{K}$. Thus one finds the picture of a blackbody radiation in the keV region, slightly modified by a hotter surface layer, which losses energy mainly through the inverse Compton effect. The case $y_0 = 20 \, \text{g cm}^{-2}$, however, allows only small contributions from the hot layer: Zel'dovich and Shakura compute the fraction η of the total energy released by the electrons through the inverse Compton effect, and find

$$\eta < 0.05 \quad \text{for} \quad \dot{M}_1, \quad \text{and} \quad \eta < 0.01 \quad \text{for} \quad \dot{M}_2.$$

Secondly, the possibility is considered of decelerating the beam of charged particles through plasma oscillations set up in the neutron star atmosphere. The mean free path may be drastically reduced in that case, and the electron temperature in the hot surface layer will rise. The electron temperature will be restricted by the fact that plasma oscillations will be strongly damped, whenever the thermal velocity of the electrons exceeds the free fall velocity. Thus T_e is limited by $10^9 \, ^\circ\text{K}$. Zel'dovich and Shakura assume $T_e = 10^9 \, ^\circ\text{K}$ throughout the y_0 layer and obtain $y_0 = 2 \, \text{g cm}^{-2}$. The temperature of the material in the interior $y \gg y_0$ is determined as in the previous case $T = (\varepsilon/a)^{1/4}$, and one obtains $T_1 = 1.1 \cdot 10^7 \, ^\circ\text{K}$, $T_2 = 6 \cdot 10^6 \, ^\circ\text{K}$ for the two different accretion rates considered. The high electron temperature produces a rise in η: $\eta = 0.96$ for $\dot{M}_1$, $\eta = 0.7$ for $\dot{M}_2$, and bremsstrahlung from the hot layer becomes significant ($50 - 100$ keV photons). The emergent spectrum will be definitely non-Planckian.

Zel'dovich and Shakura compute spectra for the 4 model cases discussed ($y_0 = 20 \, \text{g cm}^{-2}$, $\dot{M}_1$, $\dot{M}_2$; $y_0 = 2 \, \text{g cm}^{-2}$, $\dot{M}_1$, $\dot{M}_2$), and find blackbody type spectra for $y_0 = 20 \, \text{g cm}^{-2}$, and thermal spectra with a high energy tail extending above 100 keV for $y_0 = 2 \, \text{g cm}^{-2}$.

Zel'dovich and Shakura also consider the possibility of coherent radiation mechanisms in the plasma producing optical and radio waves. They attempt to fit their model for $y_0 = 2 \, \text{g cm}^{-2}$ to Sco X-1 data.

The strong dependence of the spectrum on the extension y_0 of the layer in which the incident particles are stopped indicates that the exact conditions under which accretion occurs are very important. The assumption of Zel'dovich and Shakura that plasma oscillates actually reduce the path length drastically, has to be investigated. As stated by Zel'dovich and Shakura themselves this assumption, the restriction to spherical symmetry, the neglect of the electron heat conductivity, and, most important, the neglect of magnetic field effects limit the reliability of their conclusions. Nevertheless, I believe, that any other more realistic theory of accretion should utilize their results.

It is quite clear that a strong magnetic field of the neutron star will introduce new features in the details of the accretion process. We have pointed out in Section 3 that the structure of the surface may be drastically altered by strong magnetic fields of 10^{12} to 10^{13} gauss. Instead of a comparatively thin atmosphere, the incoming matter would be stopped by solid matter of a density of $\approx 10^4$ g/cm^3. This structure, however, will again be affected by the accreting material. The properties of the incoming plasma stream depend also sensitively on the magnetic field properties. The gyration radius of 100 MeV protons in fields of 10^{12} gauss is of the order of 10^{-7} cm. It may therefore be appropriate to describe the accreting matter in a hydrodynamical approximation. Preliminary considerations (Meyer, Schmidt, personal communication) show that a wealth of interesting phenomena will occur. The magnetic field will guide the matter down, compress it to extremely high density (10^4 g/cm^3 may occur), and the material will impinge on the neutron star surface in a hot spot of very small area, and very high temperature (above 10^9 °K). The heat transfer from these tiny hot spots in the surface to larger areas is another interesting problem. The matter falling down in the magnetic field may arrange itself in bandstructures, similar to the aurora phenomena observed on earth, with the possibility that the width of these bands is of the order of the gyration radius.

The observations of pulsating X-ray sources (cf. 8.6) show that magnetic fields do indeed play an important role in the accretion process on neutron stars. So it is a realistic and challenging task to find a way to describe accretion under the influence of strong magnetic fields.

8.5. Change of the Surface Composition by Accretion

After a few years of existence neutron stars will have surfaces which consist mainly of Fe^{56} (Cameron, 1970; Rosen, 1969), and the admixture of other elements is practically negligible. Dyson (1971) has pointed out that stable chemical compounds may exist among the elements occurring in neutron star crusts. This effect may increase the percentage

Table 7. Surface redshifts, incident proton energies and gamma ray yields as functions of neutron star mass. Per proton of given energy, Q^+, $Q^{(4.43)}$ and $Q^{(6.14)}$ are the yields for positron annihilation radiation, 4.43 MeV carbon line, and 6.14 MeV oxygen line, respectively

$M/M_\odot$	Z_S	$E_p(\text{MeV})$	Q^+	$Q^{(4.43)}$	$Q^{(6.14)}$
0.37	0.053	39	0.0062	0.003	0.002
0.55	0.085	57	0.015	0.005	0.003
0.77	0.13	77	0.028	0.006	0.004
1.00	0.18	103	0.048	0.007	0.005
1.24	0.24	130	0.070	0.008	0.006
1.44	0.31	160	0.095	0.009	0.007
1.56	0.38	184	0.12	0.01	0.008
1.68	0.45	208	0.14	0.011	0.008
1.72	0.51	230	0.18	0.012	0.009

of other elements relative to Fe^{56}, but a quantitative analysis has not been done yet.

A significant modification of the surface composition will occur for neutron stars under accretion, if the accretion rate is sufficiently large. The energy of an accreted particle at the surface of a neutron star ranges from about 0.1 MeV to a few 100 MeV, depending on the neutron star mass (Table 7). Since a 100 MeV proton (1 $M_\odot$ neutron star) traverses $X_p \approx 10$ g cm^{-2} (nuclear reactions) to 100 g cm^{-2} (Coulomb collisions) of surface material before coming to rest, the amount of neutron star matter that is exposed to accreted particles is of the order

$$4\pi R^2 X_p \approx 10^{14}g \quad \text{to} \quad 10^{15}g \tag{8.20}$$

for spherically symmetric accretion, where $R = 10^6$ cm is the neutron star radius. With the accretion rate of 10^{11} g sec^{-1} from the interstellar medium, the exposed material will be turned over by accretion every 10^3 to 10^4 sec. The surface composition will thus depend more on the composition of the incoming matter, nuclear reactions on the surface, and diffusion processes into the interior, than on the original surface composition. This is even more drastic for neutron stars in binary systems, with high accretion rate.

The theoretical determination of the detailed surface composition of neutron stars affected by accretion appears to be a difficult problem, which has not yet been treated extensively. But a simplified model has been considered by Börner et al. (1973):

They consider a box of depth l and surface area 1 cm^2 at the neutron star surface continuously being hit by incoming particles of flux ϕ. The temperature is adopted as low enough so that fusion reactions are unimportant, and only spallation has to be considered. Then, assuming that a steady state has been reached, the number density $n(A, Z)$ of a spallation product (mass number A, charge Z_e) is determined by the equilibrium between diffusion and spallation of this nucleus (A, Z) and the creation of the (A, Z) nucleus through spallation of nuclei with higher mass and charge. One has

$$n(A, Z)\left(\tau_d^{-1}(A, Z) + \tau_{AZ}^{-1}\right) = \sum_i n(A', Z')_i\, \tau_{ZZ'}^{-1}, \tag{8.21}$$

where the summation on the right hand side is taken over all nuclei i with Z' or A' greater than that of the nucleus (A, Z) up to the heaviest nucleus, which is taken to be Fe^{56}. $\tau_d(A, Z)$ is the time for diffusion of the nucleus (A, Z) into the interior. The other time scales describe the spallation reactions

$$\tau_{AZ}^{-1} = \alpha\phi\sigma_{AZ}, \tag{8.22}$$

$$\tau_{ZZ'}^{-1} = \alpha\phi\sigma_{A'Z',AZ}. \tag{8.23}$$

Here σ_{AZ} is the cross section for the spallation of the nucleus (A, Z); $\sigma_{A'Z',AZ}$ is the cross section for the spallation reaction of the nucleus (A, Z), where one spallation product is the nucleus (A', Z'), $\alpha\phi$ is the flux of incoming particles interacting with nuclei. For incoming 100 MeV protons, a surface density of $1\,g\,cm^{-3}$ gives l as 10 cm (for a mean free path of $10\,g\,cm^{-2}$), and $\alpha = 0.1$. The interacting protons are not lost through the interactions with the surface material, but reappear through those same interactions. Thus the steady state equation for protons becomes

$$n(1, 1)/\tau_d(1, 1) = \alpha\phi/l \approx 10^{-2}\phi, \tag{8.24}$$

$\tau_d(1, 1)$ is the proton diffusion time. Börner et al. (1973) assumed that iron (56, 26) was the heaviest nucleus, and solved Eq. (8.21) together with (8.24) numerically in an iteration procedure. This gives the relative abundances $n(A, Z)/n(56, 26)$ as functions of ϕ, and the proton diffusion time. The diffusion times of heavier nuclei are related to the proton diffusion time by (Rosen, 1969)

$$\tau_d(A, Z)/\tau_d(1, 1) = (Z + 1)\, Z^2\, (4A)^{-1/2}. \tag{8.25}$$

The total mass density at the stellar surface was taken to be fixed at $1\,g\,cm^{-3}$. This in turn allows to obtain the absolute abundances of the elements on the neutron star surface. In Fig. 7 two results of Börner et al.

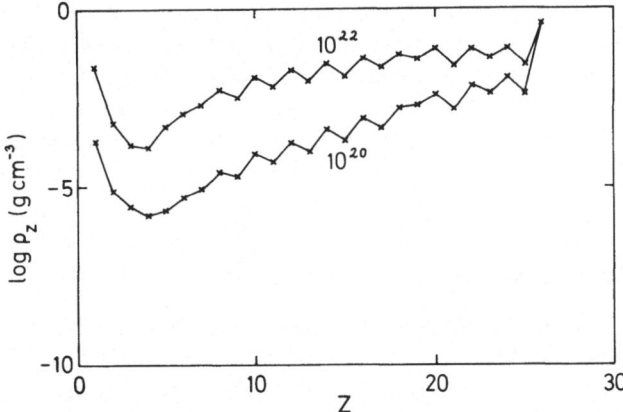

Fig. 7. Surface composition in the case of accretion. An odd-even effect is clearly present. The two parameter values $\chi = 10^{20}$, $\chi = 10^{22}$ are plotted

are reproduced. These graphs correspond to accretion from the interstellar medium, where $\phi = 5 \times 10^{21}/\mathrm{cm}^2$. Two different values for the proton diffusion time were taken, corresponding to values of χ of 10^{20} and 10^{22} ($\chi = \alpha \phi \tau_d(1, 1) = 5 \times 10^{20} \tau_d(1, 1)$). The proton diffusion time should, of course, be unambiguously determined, but in few of the uncertainties of the conventional theories, Börner et al. decided to compute the surface composition for a range of values of $\tau_d(1, 1)$.

In Fig. 7 the mass density ϱ_z of the element Z is plotted, which is obtained by adding up the densities of all isotopes of the same Z.

The abundance distribution presented in Fig. 7 may be realistic for the case of accretion of interstellar matter on old neutron stars without a magnetic field. In that case the temperature in the surface layer is practically exclusively determined by the black-body temperature corresponding to $M = 10^{11}$ g/sec. Thus $T \approx 10^5$ °K, and the only nuclear reactions possible will be spallation reactions. According to the results of the last section, there will be a hotter ($\approx 10^7$ °K), but very thin layer on the surface, but this does not allow the surface composition to be modified significantly from the one in Fig. 7.

For neutron stars in binary systems the accretion rate may be higher by a factor of 10^6, but the temperature will also increase, at least to 10^7 °K. Then the diffusion times will be much shorter too, at least by a factor 10^{-3}, and thus χ will be 10^{23} to 10^{26}. The surface composition can, however, no longer be described realistically by the simple model presented above. The temperatures in certain layers of the surface may be much higher than the black-body temperature of 10^7 °K, depending on the physics and geometry of the accretion process (magnetic field effects, concentrated hot spots etc.). Thus, as we have seen in the last section, $T = 10^9$ °K in a layer of $2 \, \text{g/cm}^2$ is present even in a grossly simplified model, and nuclear fusion reactions then become a distinct possibility, which will drastically influence the surface composition.

8.6. X-Ray Emission

The observational situation has been greatly improved during the past 2 years. The first discovery of X-ray sources was made in 1962, and until 1971 observational techniques in this field were confined to the use of sounding rockets and balloons. Then the first satellite devoted exclusively to X-ray astronomy was designed and developed by Giacconi and his colleagues at AS and E for NASA, and launched off the Kenya coast in East Africa during December 1970. The satellite, officially designated SAS–A, was christianed "Uhuru" (Swahili for freedom) in recognition of the Kenyan independence day, which coincided with the launch date. The Uhuru satellite detects X-rays in the frequency range 2 – 20 keV, and from the observations, which are supported by balloon and rocket measurements, it seems that the excitation process by which the X-rays are made favors the production of photons in this range. Fortunately the Uhuru satellite had built in the means to find temporal variations of sources, and it is in this field that the most spectacular discoveries have been made:

The binary nature of at least 2, periodically pulsating, eclipsing X-ray sources has been established by the Uhuru observations. The source found in Centaurus Cen X-3 (Giacconi et al., 1971, 1972; Schreier et al., 1972) pulsates with a period of 4.8 sec, and is seen to be eclipsing with a period of 2.087 days. The source found in Hercules Her X-1 (Tananbaum et al., 1972; Giacconi et al., 1972) has a pulsation period of 1.24 sec, and an orbital period of 1.7002 days. Her X-1 shows in addition to the occulting phenomenon a period of ≈ 36 days, where the source is present at high intensity for ≈ 9 days – pulsating and eclipsing with the 1.7 day period – and then disappears for 27 days. Similar long-term states of high and low intensity have been found in Cen X-3, but of a more irregular temporal behavior.

These beautiful experiments indicate that there has been discovered a new distinct class of sources, objects emitting pulsed X-rays and orbiting in a binary system. It seems almost inevitable to interpret the X-ray source as a magnetic neutron star, whose rotation provides the timing mechanism of the pulsations, and whose energy source for producing the X-rays is the accretion of mass from its companion in the binary system. The magnetic field should change the geometry of the accretion process sufficiently from spherical symmetry to allow the generation of pulses. For example in a strong magnetic dipole field accretion will preferably happen in the region of the magnetic poles, thus producing hot spots on the surface (Ostriker and Davidson, 1972; Lamb et al., 1972). These hot spots will radiate isotropically, but the infalling material will be more opaque to radiation propagating in the direction of infall, and less opaque perpendicular to that direction, and thus the radiation will be emitted in a cone, producing the pulse phenomen (Pringle and Rees, 1972). The pulse structure in Her X-1 is clearly a double pulse, whereas in Cen X-3 there seem to be 3 subpulses, possibly produced by 3 different hot spots.

From the short periods of the 2 X-ray sources it is clear that from the known mechanisms only rotation or oscillation of a white dwarf, or rotation of a neutron star can be employed as the time-keeping mechanism. However, recently computed white dwarf models made of carbon or carbon-burning products (Cohen et al., 1969) have – because of general relativistic effects – only stable oscillations for periods larger than 2 sec. Thus if we consider Cen X-3 and Her X-1 as two representatives of one class of objects, oscillation of white dwarfs can be ruled out as a pulse mechanism. Rotation of white dwarfs is still a possibility, but one has to go to rather extreme conditions: Listing a few dwarf models from Cohen et al. (1969)

$$\begin{array}{lccccc}
\text{Mass}/10^{33}\,\text{g} & 2.26 & 2.45 & 2.58 & 2.65 & 2.3 \\
\text{Radius}/10^{8}\,\text{cm} & 4.25 & 3.25 & 2.44 & 1.81 & 1.34,
\end{array} \tag{8.26}$$

we find that the rotation periods have to be very close to the critical one, $T = 2\pi(R^3/GM)^{1/2}$, and even then only the models close to the Chandrasekhar limit (1.33 $M_\odot$ in this case) can produce a rotation period of 1.24 sec $(M > 2.58 \cdot 10^{33}$ g). In close binary systems one would expect the rotation of the individual components to be bound to the orbital motion. They would show each other the same face all the time, just as moon and earth do. It is hard to see how in such a system one component could evolve into a white dwarf rotating close to critical velocity.

It thus seems that the compact X-ray source must be a neutron star, whose rotation provides the time-keeping mechanism. The accretion process itself is certainly drastically influenced by the rotating strong stellar magnetic field, and it is not clear to what extent the picture of spherically symmetric accretion (cf. 8.4) is applicable. As has been pointed out in Section 8.4. not much work has been done on this question. Pringle and Rees (1972) consider the physics of the accretion disk, the cloud of matter surrounding the neutron star. Davidson and Ostriker (1972) and Lamb et al. (1972) have done similar work parallel to Pringle and Rees. This interesting subject is somewhat outside the scope of the present review, which is concerned mainly with the physical processes happening on and in the neutron star. The observed spectra will depend significantly on the process very close to the neutron star, while the accretion disk has to be optically thin to keV X-rays. This means that it has a column density of $1\,\mathrm{g\,cm^{-2}}$, and absorption effects will be important only for the low frequency part of the spectrum.

The size of the accretion region depends on the structure of the magnetic field. Lamb et al. (1972) and Ostriker and Davidson (1972) consider a model, where the matter accumulates near the so-called Alfvén-surface, whose radius r_A is determined by the equilibrium between the energy density of the magnetic field and the pressure of the incoming matter

$$B^2(r_A)/8\pi = \varrho(r_A)\,v^2(r_A)\,. \tag{8.27}$$

The matter will therefore probably not accrete along field lines that close within r_A. The assumption of a dipole field (field lines: $\sin^2\theta/r = \mathrm{const}$) gives the condition that all field lines reaching the Alfvén surface have to be inside the angle

$$\sin^2\theta_c = R/r_A\,.$$

Lamb et al. (1972) then estimate the maximal size of the area on the surface where accretion occurs as

$$\pi R^2\theta_c \approx 10^{10}\ \mathrm{cm^2}\,. \tag{8.28}$$

Here values typical for a neutron star of $B = 10^{12}$ gauss, $R = 10^6$ cm, $M = 1 M_\odot$ have been employed. The accretion region thus covers at most 10^{-3} of the total surface area. This result together with the observed spectra of the X-ray sources will give restrictions for possible models.

The spectrum of Cen X-3 has been measured between 2 and 20 keV by Giacconi et al. (1971). They find a good fit to the data with either a blackbody spectrum with a temperature of about $3.3 \times 10^7\,°\mathrm{K}$ or a thermal bremsstrahlung spectrum with $kT = 15.7 \pm 1$ keV (Clark et al., 1972, point out, that taking into account the gaunt factor, which was neglected by Giacconi et al. increases kT to about 50 keV) and a cut off of 3.8 keV. The peak flux occurs at 6 keV, and at 20 keV the flux is down by a factor 50. This does not exclude the existence of a high-energy tail with a flux lower by at least 10^{-2} from the peak, as it is predicted in the simple model of Zel'dovich and Shakura (1969) discussed in 8.4. The luminosity of the source can be computed by adopting e.g. the $3.3 \times 10^7\,°\mathrm{K}$ blackbody fit, and the radiating area of (8.28)

$$\dot{E} = \sigma T^4 \times 10^{10} = 8 \times 10^{35}\ \mathrm{ergs/sec}\,. \tag{8.29}$$

Comparing this with the observed flux would place the source at a distance of slightly less than 1 kpc, whereas the suggested optical identification with WRA 795 (which is still being disputed, however) leads to a distance of 8 kpc (Börner et al., 1972). The discrepancy may be serious, but in view of the observational uncertainties is difficult to assess.

The X-ray spectrum of Her X-1 obtained from the Uhuru satellite has not been published yet. Spectral data are available from the OSO-7 observations (Clark et al., 1972). Their results indicate that Her X-1 is different from Cen X-3 insofar as it seems to produce more high frequency photons in the 15 to 30 keV range. Clark et al. (1972) claim that the data can be fitted to either a power spectrum (spectral index $\alpha = 0.6 \pm 0.15$, cut off energy 1.0 ± 0.5 keV) or a thermal bremsstrahlung spectrum ($kT = 50 \pm 20$ keV, cut off energy 1.0 ± 0.5 keV), but that they are inconsistent with a blackbody spectrum. It seems that in view of the large statistical uncertainties such a clear decision is not yet possible, and that a black-body spectrum with perhaps a high energy tail may still be a viable candidate. We have to wait for future measurements, which should eventually also enable us to discard a power-law spectrum as a likely candidate. If accretion is indeed the energy source in these objects, as seems beyond doubt now, then the observed spectrum will be of black body or thermal bremsstrahlung type, but not power law like.

To release a luminosity of 8×10^{35} ergs/sec [see (8.29)], from a typical neutron star with $\dot{E} = 0.1\, c^2 \dot{M}$, requires an accretion rate of

$$\dot{M} = 8 \times 10^{15} \text{ g/sec}. \tag{8.30}$$

This mass is accreted on an area of $\approx 10^{10}$ cm^2, and therefore

$$\dot{m} = 8 \times 10^5 \text{ g/sec cm}^2. \tag{8.31}$$

The "Eddington limit" gives for spherically symmetric accretion (8.2)

$$\dot{m}_c = 10^5 \text{ g/sec cm}^2. \tag{8.32}$$

So the $\dot{m}$ of (8.31) is above the critical accretion rate. There are, however, observational and theoretical uncertainties involved in deriving $\dot{m}$; moreover the geometrical situation is certainly not spherically symmetric, and may allow to increase the "Eddington limit" somewhat. It seems quite plausible therefore, that the neutron stars in Cen X-3 and Her X-1 are accreting mass at a constant rate, which is regulated by the "Eddington limit".

The possibility to observe neutron stars in binary systems allows in principle to determine their mass within rather narrow limits. Astronomical observations of the eclipsing system give a value for the mass

function $(M_2^3/(M_1 + M_2)^2) \sin^3 i$, where M_1 is the mass of the neutron
star, M_2 is the mass of the companion in the binary system, and $\sin i$
gives the inclination of the plane of the orbit to the line of sight ($i = 90°$
means that the line of sight is in the orbital plane). The mass function
for Her X-1 was found to be $0.85 \, M_\odot$ (Tananbaum et al., 1972), for
Cen X-3 $15 \, M_\odot$ have been measured (Schreier et al., 1972). A determination of $\sin i$, M_2 will thus give the mass of the neutron star. The value of
$\sin i$ can be obtained from measurements of the velocities of the system,
and the eclipse duration. While no definite conclusion is possible yet
for Cen X-3, it seems that $\sin i = 1$ for Her X-1 (Schmidt and Thomas,
personal communication). Furthermore Her X-1 has been identified with
the optical variable HZ Hercules, which has been known as irregular
(Liller, 1972; Bahcall and Bahcall, 1972). The light variation of $1^m.5$ of
this object (Kukarkin et al., 1970) has been shown to coincide precisely
with the exlipsing of Her X-1. Even in that case, however, the mass
M_2 of the primary is difficult to obtain within narrow limits. Almost
all the estimates for the mass of M_2 and M_1 (Wilson, 1972; Osaki, 1972;
van den Heuvel and Heise, 1972; Ruffini and Leach, 1972) rest on the
assumption that the primary star fills its critical Roche lobe. However,
Börner et al. (1972) have argued that this cannot be expected. The theory
of mass flow in binaries (see Kippenhahn and Weigert, 1967) gives a
mass loss rate for stars filling the critical Roche lobe of $3 \times 10^{-4} \, M_\odot$/year
(for $M_2 = 17 \, M_\odot$ as for the Cen X-3 system) or $10^{-6} \, M_\odot$/year (for
$M_2 = 2 M_\odot$ as for HZ Hercules). We have already pointed out that the
accretion rate onto the neutron star is probably constant in both cases
and about 10^{16} g/sec $\approx 10^{-10} \, M_\odot$/year. Since a column density of
1 g/cm^2 is sufficient to block the X-rays coming from the neutron star,
not much more than $10^{-11} \, M_\odot$ can be stored in a cloud around the
neutron star. In view of this discrepancy in the mass transfer rates
Börner et al. (1972) conclude that the primary components of Cen X-3
and Her X-1 do not fill their critical Roche surfaces.

Börner et al. (1972) point out that mass loss due to rapid rotation
and the creation of a circumstellar disk, resulting in a Be star, may be
the mechanism in the case of Cen X-3. Typical mass loss rates are around
$10^{-9} \, M_\odot$/year (Strittmatter et al., 1970). In the case of Her X-1 the large
variations of the optical star indicate that the much brighter X-ray
source is able to evaporate matter from the surface of its companion
(Schmidt, Thomas, 1973).

The attempt of Bahcall and Bahcall (1972) to determine the mass
M_2 of HZ Hercules from its spectral type is also not convincing, because
here the large variations of the optical luminosity show that HZ Hercules
is not a peaceful main sequence star. Before we will know the masses of
the neutron stars in Her X-1 and Cen X-3 with certainty, detailed models

of the systems have to be constructed in the hope that the observations will bring enough restrictions to single out the correct one.

Several other fluctuating, but not periodically pulsating X-ray sources have been observed (for a review see Gursky, 1972). Accretion of mass from the companion in a binary system onto a neutron star or black hole is probably the energy source in all cases. Although a neutron star under accretion will eventually collect enough mass to evolve beyond the maximum stable mass configuration and collapse into a black hole, no absolutely convincing case for a black hole has been made for any of these objects.

The X-rays emitted from old neutron stars (defunct pulsars) accreting interstellar matter may also be detectable. Ostriker et al. (1970) and Silk and Weinberg (1972) have in fact suggested that the low-energy (250 eV) diffuse X-ray emission from the galactic disk could be due to accretion on neutron stars from the interstellar medium. Assuming an accretion rate of 10^{11} g/sec per neutron star gives the peak flux at a few 100 eV. In order to account for the observations, this mechanism requires a neutron star density in interstellar space of $\approx 3 \times 10^{-2}\,\mathrm{pc}^{-3}$. Since soft X-rays are absorbed on distances of $\approx 100\,\mathrm{pc}$ in the interstellar medium, the observation of the 250 eV peak gives information only on nearby sources, and this neutron star density represents the density of such objects in the vicinity of the sun.

It should be noted that Strittmatter et al. (1972) and Gorenstein and Tucker (1972) have pointed out difficulties with the accretion model on neutron stars as an explanation for the galactic soft X-ray background. If these difficulties are indeed real, then the local density of 3×10^{-2} neutron stars per pc^3 should be considered as an upper limit only.

8.7. Gamma Radiation Emission

The satellite SAS-B, which was launched by NASA in December 1972, is devoted exclusively to gamma radiation observations. Will it bring as dramatic an increase of our knowledge as the Uhuru satellite? The balloon observations available to date have problems with the strong background, which has prevented the detection of gamma ray sources with a flux of less than $10^{-3}\,\mathrm{cm}^{-2}\,\mathrm{sec}^{-1}$.

It is quite clear from theoretical considerations that gamma ray emission has to be expected from neutron stars under accretion. Shvartsman (1972) tries to enumerate some of the possibilities:

a) Meson creation is only important on neutron stars with very high gravitational potential such that the energy of the incident protons is greater than 120 MeV, and hence the gamma ray production by the ensuing decay of these mesons is negligible except for the most massive neutron stars (cf. Table 7). The most efficient mechanism is the decay $\pi^0 \to 2\gamma$, and the gamma ray luminosity (35 – 60 MeV gamma rays) may according to Shvartsman's estimates amount to 10^{-4} of the total luminosity in the highest mass models. However, for a typical neutron star of 1 $M_\odot$ the flux is already down to $\cong 10^{-8}$ of the total luminosity. It seems therefore that with present equipment there is no chance to detect gamma rays from meson decay.

  b) *Neutron capture* by protons to form deuterium may occur. A gamma ray line at 2.2 MeV (redshifted by the gravitational potential of the star) is emitted, and Shvartsman (1972) estimates the flux to be about 10^{-5} of the total luminosity. Even for the strong X-ray sources which emit 10^{36} ergs/sec this would amount to a flux of only 10^{-6} photons/cm² sec.

  c) *Inelastic scattering* of protons by nuclei will give rise to the emission of gamma photons too. Shvartsman (1972) considers some reactions involving the nuclei C^{12}, N^{14}, O^{16}, and Ne^{20}; the emitted photons have energies between 2 and 9 MeV, and the flux is again about 10^{-5} of the total luminosity. Shvartsman assumes these nuclei to be present in an abundance of 0.1 %. To obtain a reliable estimate, however, one will have to compute the surface composition too.

  d) *Thermonuclear reactions* in the hot surface layer (Zel'dovich and Shakura, 1969) produce, according to Shvartsman's estimate, gamma rays in the range from 10 – 20 MeV with a flux of less than 10^{-6} of the total luminosity.

  These calculations give a rough idea of what to expect, but it seems prudent to accept Shvartsman's (1972) quantitative analysis with caution. He discusses by no means all the possible gamma radiation producing reactions, and he relies heavily on the model of Zel'dovich and Shakura (1969), which has been discussed in Section (8.4.). As has been mentioned their model of spherically symmetric accretion does not apply to the binary X-ray sources, and the actual ranges of particles as well as the temperature distribution in the accretion region may be drastically different. Therefore it is difficult to judge whether or not Shvartsman's calculations give reliable estimates of the gamma ray fluxes to be expected. He is probably too optimistic what regards the identification of gamma ray sources.

  The electron positron annihilation radiation at 511 keV is treated by Shvartsman (1972) only in the context of the decay $\mu^+ \to \pi^+ \to e^+$. He thus arrives at very low fluxes (10^{-10} for a 1 $M_\odot$ star).

  But recently Johnson et al. (1972) have reported low energy gamma ray observations from the galactic center region. These observations provide evidence for a statistically significant spectral feature at 473±30 keV with a total photon flux in the feature of $\sim 1.8 \times 10^{-3}\,\mathrm{cm}^{-2}\,\mathrm{sec}^{-1}$. Ramaty et al. (1973) have interpreted this line emission as gravitationally redshifted positron annihilation radiation from the surface of old neutron stars, which are accreting interstellar matter. Ramaty et al. (1973) have calculated positron production by protons slowing down and coming to rest in material consisting solely of carbon, nitrogen and oxygen in ratios by number given by $C:N:O = 0.5:0.1:1$. The positron yield Q^+ per proton as evaluated by Ramaty et al. is presented in Table 7 as a function of the incident energy E_p at the surface of the neutron star. E_p as well as the surface redshift Z_S are also given in Table 7 as functions of the neutron star mass.

The observed line feature at $473 \pm 30\,\mathrm{keV}$ is consistent with redshifts ranging from ~ 0.016 to 0.13. An average value of $Z_S = 0.1$ corresponds according to Table 7 to $M/M_\odot \cong 0.6$, $E_p \cong 65$ MeV and $Q^+ \cong 0.02$. Ramaty et al.'s interpretation of the observations has as a consequence that the majority of these old neutron stars have masses of less than $0.8\ M_\odot$.

For an accretion rate of 10^{11} g/sec or 6×10^{34} particles/sec the redshifted positron annihilation radiation yield of a neutron star is $6 \times 10^{34} \cdot Q^+$ photons/sec (10^{33} photons/sec for $M = 0.6\ M_\odot$). This result is roughly in agreement with estimates derived from the calculation of the surface composition by Börner et al. (1973). But if the observed line feature at 473 keV comes entirely from a region around the galactic center, a total of 1.5×10^{10} neutron stars are required in the field of view of the detector at an average distance of 10 kpc. For a volume of 1.5×10^{10} pc^3 (opening angle of $24°$ in both latitude and longitude of the gamma ray telescope, and a disk thickness of 1 kpc) the neutron star density is $1\,\mathrm{pc}^{-3}$.

This number is rather large, but a higher accretion rate would, of course, lead to an increased emissivity per neutron star. Ramaty et al. (1973) point out that the accretion rate per neutron star cannot be arbitrarily increased because of limitations imposed by X-ray observations. The kinetic energy of the accreted particles must be released, and therefore the observed flux of 1.8×10^{-3} photons/cm^2 sec at 473 keV requires the dissipation of 10^{41} ergs/sec at the galactic center. X-ray observations in the keV regime rule out such a large luminosity, and therefore the 10^{41} ergs/sec have to be emitted as ultraviolet or soft X-rays, which are absorbed in the interstellar medium and cannot be observed at earth. A neutron star radiating as a black body with kT less than 500 eV must have an accretion rate of less than $\cong 10^{16}$ g/sec. Thus Ramaty et al. (1973) show that if the redshifted annihilation radiation is produced on neutron stars in binary systems at least 10^5 neutron stars are required to account for the observations.

The model of Ramaty et al. (1973) presents an interesting interpretation of an exciting measurement. Future observations are needed to
show whether the measurement of Johnson et al. (1972) recorded a true
effect or not. But not withstanding the caution with which one has to
regard any present day gamma radiation observation, the excellent
diagnostic possibilities in observing gamma ray lines from neutron
stars are quite clear already from this first, and not yet well-established
result: The redshift of the positron annihilation line (which should
always be the strongest line present) can be measured directly, and then
all the physical parameters of the neutron star emitting the radiation
are known, or at least, as in the observations of Johnson et al. (1972)
the statistical distribution of the parameters can be found.

A single nearby neutron star at a distance of 5 pc accreting interstellar matter (10^{11} g/sec) would emit a redshifted 511 keV line with
a flux at earth of $\cong 10^{-6}$ photons/cm^2 sec. This object may be completely
dead in any other way, but we still could hope to observe it in the 511 keV
line with equipment about 100 times as sensitive as the present available.

A similar flux of $\cong 10^{-6}$ photons/cm^2 sec at 511 keV might be
expected from the pulsating X-ray sources Her X-1 and Cen X-3. The
number rests on the assumption that the radiation yield at 511 keV
increases proportional to the accretion rate, an assumption which has
to be investigated in detailed models. But if we believe this number, the
chances for detection are very good. The sensitivity of the detectors is
probably greatly increased, because the gamma ray line will appear
pulsed with precisely the period and phase of the X-ray pulsations. We
may thus be able to use gamma ray spectroscopy in observing the pulsating X-ray sources in the near future, which would help in understanding
the intricate details of the accretion process on neutron stars.

In conclusion I would like to point out that accretion on black holes
does not produce gamma rays, and thus observation of gamma radiation
from any X-ray source will clearly establish its neutron star nature.

Acknowledgements. I would like to take this opportunity to express special gratitude
to Hans Bethe (Ithaca), Al Cameron (New York), Jeffrey Cohen (Philadelphia), Ludwig
Biermann, Peter Kafka, Friedrich Meyer, and H. U. Schmidt (Munich).

References

Ambartsumyan, V. A., Saakyan, G. S.: Soviet Astron. AJ **4**, 187 (1960).
Anderson, P. W., Palmer, R. G.: Nature Phys. Sci. **281**, 145 (1970).
Bahcall, J. N., Bahcall, N.: IAU Circ. No. 2426, 1972.
Bahcall, J. N., Bahcall, N.: Preprint (Princeton & Tel Aviv) (1972).
Baldwin, J. E.: Proc. IAU Symp. No. **46**, 22 (1971).
Banerjee, B., Chitre, S. M., Garde, V. K.: Phys. Rev. Letters **25**, 1125 (1970).

Barkat, Z., Buchler, J.-R.: Astrophys. Letters 7, 167 (1971).
Baym, G., Pethick, C., Pines, D., Ruderman, M.: Nature 224, 872 (1969).
Baym, G., Pethick, C., Sutherland, P.: Astrophys. J. 170, 299 (1971 a).
Baym, G., Bethe, H. A., Pethick, C.: Nucl. Phys. A 175, 225 (1971 b).
Baym, G., Pines, D.: Ann. Phys. 66, 000 (1971).
Bethe, H. A.: Personal Communication (1969).
Bethe, H. A.: Ann. Rev. Nucl. Sci. 21, 000 (1971).
Bethe, H. A., Börner, G., Sato, K.: Astron. Astrophys. 7, 279 (1970).
Bethe, H. A., Johnson, M.: IAU Symp. no. 53 (in press) (1973).
Bonazzola, S., Ruffini, R.: Phys. Rev. 187, 1767 (1969).
Bondi, H.: Proc. Roy. Soc. A 281, 39 (1964).
Börner, G., Cohen, J. M.: Nature Phys. Sci. 232, 30 (1971).
Börner, G., Sato, K.: Astrophys. Space Sci. 12, 40 (1971).
Börner, G., Cohen, J. M.: Astron. Astrophys. 19, 109 (1972 a).
Börner, G., Cohen, J. M.: In preparation ("Pulsar Speed-Ups") (1972 b).
Börner, G., Meyer, F., Schmidt, H. U., Thomas, H.-C.: Proc. DAG meeting, Wien (1972).
Börner, G., Cohen, J. M.: Astrophys. J. (in press) (1973).
Börner, G., Ramaty, R., Tsuruta, S.: In preparation (1973).
Boynton, P., Groth, E., Partridge, P., Wilkinson, D.: IAU Circ. No. 2179, 1969.
Brill, D. R., Cohen, J. M.: Phys. Rev. 143, 1011 (1966).
Buchler, J.-R., Ingber, L.: Properties of the neutron gas, OAP-238, 1971.
Cameron, A. G. W.: Ann. Rev. Astron. Astrophys. 8, 179 (1970).
Cameron, A. G. W., Mok, M.: Nature 215, 464 (1967).
Canuto, V., Chitre, S. M.: IAU Symp. no. 53 (in press) (1973).
Clark, G.W., Brodt, H.V., Lewin, W.H.G., Markert, T.H., Schnopper, H.W., Sprott, G.F.: Astrophys. J. 177, L 109 (1972).
Clark, J. W., Chao, N. C.: Nature Phys. Sci. 281, 145 (1972).
Clark, J. W., Chao, N. C., Yang, C. H.: Nucl. Phys. A 179, 320 (1972).
Cohen, J. M., Relativity theory and astrophysics (ed. J. Ehlers) A.M.S. (Providence, R.I.) (1965).
Cohen, J. M., Brill, D. R.: Nuovo Cimento 50 B, 209 (1968).
Cohen, J. M., Cameron, A. G. W.: Astrophys. Space Sci. 10, 227 (1971).
Cohen, J. M., Langer, W. D., Rosen, L. C., Cameron, A. G. W.: Astrophys. Space Sci. 5, 213 (1969).
Cohen, J. M., Lapidus, A., Cameron, A. G. W.: Astrophys. Space Sci. 5, 113 (1969).
Cohen, J. M., Toton, E. T.: Astrophys. Letters 7, 213 (1971).
Davidson, K., Ostriker, J. P.: Princeton Univ. preprint, 1972.
Day, B. D.: Rev. Mod. Phys. 39, 719 (1967).
Dyson, F.: Ann. Phys. 63, 1 (1971).
Eddington, A. S., The internal constitution of the stars. Cambridge Univ. Press 1926.
Feynman, R. P., Metropolis, N., Teller, E.: Phys. Rev. 75, 1561 (1949).
Finzi, A., Wolf, R. A.: Astrophys. J. 155, L 107 (1969).
Frautschi, S., Bahcall, J. N., Steigman, G., Wheeler, J. C.: Comm. Astrophys. Space Phys. 3, 121 (1971).
Giacconi, R., Gursky, H., Kellogg, E., Schreier, E., Tananbaum, H.: Astrophys. J. 167, L 67 (1971).
Giacconi, R., Murray, S., Gursky, H., Kellogg, E. M., Schreier, E., Tananbaum, H.: Astrophys. J. 178, 281 (1972).
Gold, T.: Nature 218, 731 (1968).
Goldreich, P., Julian, W. H.: Astrophys. J. 157, 869 (1969).
Gorenstein, P., Tucker, W. H.: Astrophys. J. 176, 333 (1972).
Groth, E.: Ph. D. Thesis. Princeton University 1971.

Gunn, J. E., Ostriker, J. P.: Ap. J. **157**, 1395 (1969).
Gursky, H.: Les Houches lectures (in press) (1972).
Hagedorn, R.: Nuovo Cimento Suppl. **6**, 311 (1968).
Harrison, B. K., Thorne, K. S., Wakano, M., Wheeler, J. A.: Gravitation theory. Chicago Univ. Press 1964.
Hewish, A., Bell, S. J., Pilkington, J. D., Scott, P. F., Collins, R. A.: Nature **217**, 709 (1968).
Hoffberg, M., Glassgold, A. E., Richardson, R. W., Ruderman, M.: Phys. Rev. Letters **24**, 775 (1970).
Johnson III, W. N., Harnden, F. R., Haymes, R. C.: Astrophys. J. **172**, L 1 (1972).
Kaplan, J. I., Glasser, M. L.: Phys. Rev. Letters **28**, 1977 (1972).
Kippenhahn, R., Weigert, A.: Z. Astrophys. **65**, 251 (1967).
Kukarkin, B. V., Parenago, P. P., Efremov, Y. I., Kholopov, P. N.: General catalog of variable stars (Moscow, Acad. Sci.) (1969).
Lamb, F. K., Pethick, C. J., Pines, D.: Univ. Illinois preprint (1972).
Landau, L.: Phys. Z. Sovjetunion **1**, 285 (1932).
Langer, W. D., Rosen, L. C.: Ap. Space Sci. **6**, 217 (1970).
Leach, R. W., Ruffini, R.: Preprint (1972).
Leung, Y. C., Wang, C. G.: Astrophys. J. **170**, 499 (1971).
Levinger, J. S., Simmons, L. M.: Phys. Rev. **124**, 916 (1961).
Liller, W.: IAU Circ. No. 2415 and 2427 (1972).
Lohsen, E.: Nature Phys. Sci. **236**, 71 (1972).
Manchester, R. N., Taylor, J. H.: Astrophys. Letters **10**, 67 (1967).
Maran, S. P., Cameron, A. G. W.: Earth Extraterr. Sci. **1**, 3 (1969).
Melrose, D. B., Cameron, A. G. W.: Canad. J. Phys. **46**, S 472 (1968).
Meyer, F.: Festschrift für C. F. von Weizsäcker (in press) (1972).
Milne, D. K.: Aust. J. Phys. **23**, 425 (1970).
Mueller, R. O., Rau, A. R. P., Spruch, L.: Nature Phys. Sci. **234**, 31. (1971).
Myers, W. D., Swiatecki, W. J.: Nucl. Phys. **81**, 1 (1966).
Negele, J. W.: Proc. IAU Symp. No. 53 (in press) (1973).
Nemeth, J., Sprung, D. W. L.: Phys. Rev. **176**, 1496 (1968).
Oppenheimer, J. R., Volkoff, G. M.: Phys. Rev. **55**, 374 (1939).
Osaki, Y.: Publ. Astron. Soc, Japan **24**, 419 (1972).
Ostriker, J. P., Rees, M. J., Silk, J.: Astrophys. Letters **6**, 179 (1970).
Pandharipande, V. R.: Nucl. Phys. A **174**, 641 (1971 a).
Pandharipande, V. R.: Hyperonic matter. Tata Institute preprint 1971.
Pandharipande, V. R.: IAU Symp. no. 53 (in press) (1973).
Pines, D., Shaham, J., Ruderman, M.: IAU Symp. no. 53 (in press) (1973).
Prendergast, K. H., Burbidge, G. R.: Astrophys. J. **151**, L 83 (1968).
Pringle, J. E., Rees, M. J.: Astron. Ap. **21**, 1 (1972).
Radhakrishnan, V., Manchester, R. N.: Nature **222**, 228 (1969).
Rajaraman, R., Bethe, H. A.: Rev. Mod. Phys. **39**, 745 (1967).
Ramaty, R., Börner, G., Cohen, J. M.: Astrophys. J. (in press) (1973).
Ravenhall, D. G., Bennett, C. D., Pethick, C. J.: Phys. Rev. Letters **28**, 978 (1972).
Rees, M., Trimble, V. L.: Astrophys. Letters **5**, 93 (1970).
Rees, M., Trimble, V. L., Cohen, J. M.: Nature **229**, 395 (1971).
Reichley, P. E., Downs, G. D.: Nature **222**, 229 (1969).
Reichley, P. E., Downs, G. D.: Nature Phys. Sci. **234**, 48 (1971).
Reid, R.: Ann. Phys. **50**, 411 (1968).
Richards, D., Pettengill, G., Roberts, J., Counselman, C., Rankin, J.: IAU Circ. No. 2181 (1969).
Rosen, L. C.: Astrophys. Space Sci. **1**, 372 (1968).
Rosen, L. C.: Astrophys. Space Sci. **5**, 150 (1969).

Ruderman, M.: Nature **223**, 597 (1969).
Ruderman, M.: Phys. Rev. Letters **27**, 1306 (1971).
Sawyer, R. F.: Phys. Rev. Letters. **29**, 382 (1971).
Sawyer, R. F., Scalapino, D. J.: preprint – UC Santa Barbara (1972).
Scalapino, D. J.: Phys. Rev. Letters **29**, 386 (1972).
Schmidt, H. U., Thomas, H.-C.: In preparation (1973).
Schreier, E., Levinson, R., Gursky, H., Kellogg, E., Tananbaum, H., Giacconi, R.: Astrophys. J. **172**, L 79 (1972).
Shklovsky, I.: Astrophys. J. **150**, L 48 (1967).
Shvartsman, V. F.: Radiofisika **13**, 1852 (1970).
Shvartsman, V. F.: Soviet Astron. A J **15**, 342 (1971).
Shvartsman, V. F.: Astrophysics **6**, 56 (1972).
Siemens, P. J.: Nucl. Phys. A **141**, 225 (1970).
Siemens, P. J., Pandharipande, V. R.: Nucl. Phys. A **173**, 561 (1971).
Silk, J., Weinberg, S. L.: Astrophys. J. **175**, L 29 (1972).
Strittmatter, P., Brecher, K., Burbidge, G.: Astrophys. J. **179**, 91 (1972).
Strittmatter, P., Robertson, J. W., Faulkner, D. J.: Astron. Astrophys. **5**, 426 (1970).
Tananbaum, H., Gursky, H., Kellogg, E., Levinson, R., Schreier, E., Giacconi, R.: Astrophys. J. **174**, L 143 (1972).
Thorne, K. S., Novikov, I. D.: OAP-304 (1973).
Tolman, R. C.: Phys. Rev. **56**, 364 (1939).
Trimble, V.: Astron. J. **73**, 535 (1968).
Trimble, V., Woltjer, L.: Astrophys. J. **163**, L 97 (1971).
Tsuruta, S., Cameron, A. G. W.: Can. J. Phys. **44**, 1863 (1966).
Tucker, W. H.: Astrophys. J. **167**, L 85 (1971).
van den Heuvel, E. P. J., Heise, J. H.: Nature Phys. Sci. **239**, 67 (1972).
Wallerstein, G., Silk, J.: Astrophys. J. **170**, 289 (1971).
Weiss, R. A., Cameron, A. G. W.: Can. J. Phys. **47**, 2171 (1969).
Wheeler, J. A.: Ann. Rev. Astron. Ap. **4**, 423 (1966).
Wheeler, J. C.: Astrophys. J. **169**, 105 (1971).
Wilson, R. E.: Astrophys. J. **174**, L 27 (1972).
Woltjer, L.: B.A.N. **14**, 39 (1958).
Zel'dovich, Ya. B.: Soviet Phys. Dokl. **9**, 195 (1964).
Zel'dovich, Ya. B., Novikov, I. D.: Dokl. Akad. Nauk SSR, **158**, 811 (1964).
Zel'dovich, Ya. B., Novikov, I. D.: Rel. Astrophys. Vol. 1 (Chicago Univ. Press) (1971).
Zel'dovich, Ya. B., Shakura, N. I.: Soviet Astron.-AJ **13**, 175 (1969).

Dr. Gerhard Börner
NASA/Goddard Space Flight Center
Greenbelt, Maryland 20771 USA
and
Max-Planck-Institut für Physik und Astrophysik
D-8000 München 40
Föhringer Ring 6
Federal Republic of Germany
(present address)

Black Holes: The Outside Story

John Stewart and Martin Walker

Contents

Notation: Although there is no universally accepted system of notation in gravitation theory or in astrophysics, we have tried as far as possible to conform to general usage. We note here some salient points with which a general reader may not be familiar.

Since we have occasion to use a $1 + 3$ timelike-spacelike decomposition, we use the signature $(- + + +)$ for the metric tensor. Tensor indices are denoted $i, j, k, \ldots$; tetrad indices $a, b, c, \ldots$ In a $1 + 3$ split, $\alpha, \beta, \gamma, \ldots$ denote indices of spacelike quantities. The covariant derivative is written ∇_i; directional derivatives such as $u^i \nabla_i$ are occasionally written D or ∇_u. Square brackets denote skew-symmetrization and round brackets symmetrization operations on the indices enclosed. When applied to differential forms (usually denoted ω) as in Section 4.2, d denotes exterior differentiation. In Chapters 4 and 5 especially, use is made of the Newman-Penrose [29] formalism.

Introduction

The object of this review is to discuss a number of properties of the space-time region outside a black hole within the context of General

Relativity. The model chosen to describe this region is (part of) Kerr's solution [1] of Einstein's vacuum gravitational field equations. Section 1 sketches the arguments which lead one to suppose that Kerr's spacetime is relevant to such a discussion. In Section 2, a brief account is given of those features of Kerr's spacetime which are germane to the questions considered in the remaining sections, in particular the symmetries and related coordinate system found by Boyer and Lindquist [2], the algebraic type of the field [3], a certain preferred tetrad system, "locally non-rotating frames", introduced by Bardeen [4], the event horizon [5, 6, 21], and the ergosphere [6, 7, 8, 64]. In Section 3 a description is given of all timelike and null geodesics (test particle and photon trajectories) in the exterior region [2, 9, 10, 11, 12, 13, 61]. A discussion of the tidal accelerations experienced by bodies orbiting in the equatorial plane of a Kerr black hole [14] and of the corresponding Roche problem [15] is presented in Section 4. Finally Section 5 contains an account of how and why one can set up equations for discussing scalar [16], electromagnetic [17], and gravitational [18] perturbations of the Kerr field, and in particular shows why the equations decouple.

1. Motivation

It is by now well known that a star whose mass is much greater than that of the sun cannot remain in equilibrium once its thermonuclear energy sources have been exhausted, and must undergo gravitational collapse. According to present theories, gravitational collapse can result in only one of three stable, final states [71]: a white dwarf [68, 69], a neutron star [70, 81], or a black hole. During the early stages of the collapse, the star might eject matter in such a way that each resulting component would have a sufficiently small mass to remain stable as a white dwarf or neutron star. If this does not happen, further collapse is inevitable. Under fairly general conditions [72], it seems that *trapped surfaces* will form. A trapped surface [5, 6, 58] is a surface surrounding the collapsing matter having the property that even the light emitted from it in an outward direction converges to the future. The development of a trapped surface signifies a point of no return for collapse, since rigorous theorems [19, 20] then imply the existence of a space-time *singularity*, and the existence of an *absolute event horizon* [5, 6, 21]. Here the term singularity does not necessarily mean that curvature becomes unbounded, but that space-time must be geodesically incomplete [65], i.e. free fall orbits cannot be extended indefinitely. In known exact models of collapse however, the incompleteness is due to infinite curvature. The event horizon is a null hypersurface surrounding the collapsing star, which is absolute in the sense that it represents the *boundary* of the exterior region from which particles or photons can escape to infinity. Since light cannot now escape from the trapped star, such a state is called a *black hole*.

If trapped surfaces do not form during collapse, there are no theorems which assert the existence or otherwise of either singularities or event horizons [83]. In the case of a black hole, the potentially embarrassing singularity cannot influence events in the outside world because the event horizon prevents information from getting out. One can ask however, whether another type of singularity might not evolve, a naked singularity, which is not hidden behind an event horizon. A simple example is a Kerr spacetime with angular momentum parameter larger than the mass. Attempts to construct models in which a naked singularity evolves from regular initial conditions have so far failed [22] however [1]. It has therefore been suggested that there might be a cosmic censor [6] who requires all singularities to be decently clothed by an event horizon. In this case, a black hole is inevitably the end point of catastrophic gravitational collapse. This is the point of view we shall adopt.

Suppose then that all of the dynamical processes associated with the collapse have ended and that the resulting black hole has settled down to a stationary state. We idealize somewhat and assume further that space-time outside the event horizon is empty and asymptotically flat, and that the star has no residual charge. Such a description should provide a reasonable approximation to the state of a collapsed star provided the star is sufficiently isolated from other objects and sufficient time has elapsed since the collapse began. We therefore ask how many stationary, asymptotically flat, vacuum black holes gravitational theory has to offer.

Surprisingly, perhaps, there is essentially only one [23]. The word "essentially" is a double hedge. Firstly, there is actually a two parameter family of such black hole spacetimes readily available, namely that of Kerr [1], the parameters being the mass M and angular momentum per unit mass a with [2] $a < M$. Secondly it has not been possible up to the present time to exclude the existence of other two-parameter families disjoint from, but not continuously deformable into, the Kerr family [24]. Since these other hypothetical families could not be continuously deformed into the spherically symmetric Schwarzschild spacetime either,

---

[1] Yodzis, Seifert and Müller zum Hagen [82] have recently shown that spherically symmetric collapse can lead to singularities which are not hidden within black holes. It remains to be seen, however, whether this can be achieved with realistic matter and a realistic equation of state.

[2] Mathematically speaking, M is not really a parameter since the metric can always be rescaled to set $M = 1$.

however, one does not regard their potential existence as especially threatening. Thus there are compelling theoretical grounds for thoroughly understanding the Kerr family of spacetimes.

Further motivation for studying the Kerr fields comes from astrophysical arguments concerning the evolution of cores of galaxies. Lynden-Bell [25] and Lynden-Bell and Rees [63] argued that one might expect to find black holes with masses in the range $10^7 - 10^9$ solar masses commonly occurring in the centres of galaxies. Bardeen [26] has pointed out that accretion onto these black holes from the orbiting matter in the galactic disk would produce a Kerr black hole, with angular momentum parameter only slightly smaller than the mass, after a finite amount of material had been accreted.

2. Coordinate System and Tetrads

2.1. The Boyer-Lindquist Coordinate System

Since we are concerned here only with what goes on outside a black hole, we shall generally understand the Kerr spacetime to be the (geodesically incomplete) asymptotically flat region external to the event horizon. With this proviso, the Kerr spacetime is stationary and axially symmetric, or equivalently [27], invariant under a 2-parameter abelian isometry group with 2-dimensional timelike orbits. Let ξ^i denote the unique (stationary) Killing vector whose norm at infinity is -1, and let ζ^i denote the unique (axi-symmetry) Killing vector whose orbits are spacelike circles, and whose canonical parameter ranges through the real numbers modulo 2π. Since ξ^i and ζ^i commute, the dual of $\xi^{[i}\zeta^{j]}$ is proportional to the skew product of two gradients dt, $d\varphi$. The scalars t, φ can be fixed uniquely up to additive constants by requiring $\xi^i V_i t = \zeta^i V_i \varphi = 1$, $\xi^i V_i \varphi = \zeta^i V_i t = 0$; then dt is timelike, $d\varphi$ spacelike, and t and φ can be chosen as two coordinates in that region in which the 2-dimensional group orbits are timelike. Now the action of the group is orthogonally transitive [28], i.e. the group orbits are orthogonal to 2-dimensional spacelike submanifolds. Choose one of these and project one of the repeated principal null congruences [1, 3] onto this surface. The projection will give a family of lines on the surface; let x be a function whose gradient (in the surface) is orthogonal to the lines, and let y be a parameter along the lines. Then define x and y throughout the rest of spacetime by demanding that they be constant along group orbits. Actually, this construction only determines x and y up to functions of themselves: $x \to \theta(x)$, $y \to r(y)$. The function r can be fixed up to a constant factor by noting that it can be chosen to be an affine parameter along the chosen repeated principal null congruence [1, 2, 3]. The function θ

can be fixed by requiring, for example, that the twist of the chosen repeated principal null congruence be proportional to $\cos\theta$. These four scalars t, r, θ, φ are the Boyer-Lindquist (BL) coordinates [2], and with respect to them the Kerr metric takes the form [3]

$$\mathrm{d}s^2 = -(1 - 2Mr/\Sigma)\,\mathrm{d}t^2 - 2(2Mr/\Sigma)\,a\,\sin^2\theta\,\mathrm{d}t\,\mathrm{d}\varphi$$
$$+ (\Sigma/\Delta)\,\mathrm{d}r^2 + \Sigma\,\mathrm{d}\theta^2 + (\mathscr{A}/\Sigma)\sin^2\theta\,\mathrm{d}\varphi^2\,, \tag{2.1.1}$$

where

$$\Sigma := r^2 + a^2\cos^2\theta$$
$$\Delta := r^2 + a^2 - 2Mr \tag{2.1.2}$$
$$\mathscr{A} := (r^2 + a^2)^2 - \Delta a^2\sin^2\theta\,,$$

and a and M are real parameters with $0 \leq |a| \leq M$. As is well-known [2, 10], the event horizon is located at the larger root $r_+ := M + (M^2 - a^2)^{1/2}$ of the equation $\Delta = 0$. The region of the Kerr spacetime in which we are interested is then covered completely by the above coordinates, apart from the usual difficulty [24] with the coordinate φ which is undefined on the symmetry axis $\theta = 0, \pi$, with ranges

$$-\infty < t < \infty$$

$$r_+ < r < \infty$$

$$0 \leq \theta \leq \pi$$

$$0 \leq \varphi \leq 2\pi\,.$$

We shall work exclusively in this coordinate system in the following.

A convenient orthonormal basis of 1-forms for the Kerr metric can be obtained by rewriting Eq. (2.1.1) in the form

$$\mathrm{d}s^2 = -[\alpha\,\mathrm{d}t]^2 + [\mu\,\mathrm{d}r]^2 + [\Sigma^{1/2}\,\mathrm{d}\theta]^2 + [\beta(\mathrm{d}\varphi - \omega\,\mathrm{d}t)]^2$$
$$= -(\omega^t)^2 + (\omega^r)^2 + (\omega^\theta)^2 + (\omega^\varphi)^2\,. \tag{2.1.3}$$

Observe that $\alpha = (1 - 2Mr/\Sigma)^{1/2}$ is the norm of the stationary Killing vector $\xi^i = \delta^i_t$, $\beta = [(\mathscr{A}/\Sigma)\sin^2\theta]^{1/2}$ is the norm of the spacelike Killing

---

[3] The sign of the parameter a has been changed from that of Eq. (2.13) of Boyer and Lindquist [2]. See Section 2.2.

vector $\zeta^i = \delta^i_\varphi$, $\omega = -(2Mr/\Sigma)\, a \sin^2\theta$ is their inner product, $\mu = (\Sigma/\Delta)^{1/2}$, and

$$\mathscr{A}/\Sigma = r^2 + a^2 + (2Mr/\Sigma)\, a^2 \sin^2\theta\,.$$

It will prove useful in the discussion of tidal accelerations in the Kerr field to make use of the algebraic structure of its Weyl curvature tensor, which, since we have a vacuum spacetime, is the Riemann curvature tensor. The Weyl tensor has a pair of repeated principal null directions tangent to nonshearing, geodesic null congruences. The advantage of this fact when performing computations involving the curvature tensor is, that in a Newman-Penrose (NP) [29] null tetrad which incorporates the repeated principal null directions (RPND's), only a single tetrad component of the Weyl tensor fails to vanish. The appropriate tetrad has been given by Kinnersley [30]. In BL coordinates it is

$$\begin{aligned}
l^i &= \tfrac{1}{2}\,\Sigma^{-1}(r^2 + a^2, \Delta, 0, a) \\
n^i &= \Delta^{-1}(r^2 + a^2, -\Delta, 0, a) \\
m^i &= (2\Sigma)^{-1/2}\,(ia\sin\theta, 0, 1, i/\sin\theta)
\end{aligned} \qquad (2.1.4)$$

normalized so that [4]

$$g_{ij} = -2l_{(i}n_{j)} + 2m_{(i}\bar{m}_{j)}\,, \qquad (2.1.5)$$

with the sole nonvanishing curvature component

$$\Psi_2 = C_{ijkl}\,l^i m^j n^k \bar{m}^l = -M(r - ia\cos\theta)^{-3}\,. \qquad (2.1.6)$$

Observe that both l^i and n^i are future-pointing, and that l^i is outgoing (in the sense that its r-component is increasing to the future), while n^i is ingoing. Equation (2.1.4) differs from Kinnerley's tetrad in that we have taken n^i rather than l^i to be affinely parametrized. This ensures that in a coordinate system which regularly covers the future branch of the event horizon, the components of the vector l^i, which generates the horizon, remain finite. This feature is important when computing tidal accelerations for orbits "near" the horizon in the $a = M$ case (see 4.2). It is perhaps also worth mentioning that Ψ_2 is a scalar invariant of the curvature; with (2.1.6), this shows that Kerr's spacetime admits no additional symmetries.

We will have occasion to make use of two other tetrad systems for the Kerr metric, but will postpone one of them (Bardeen's locally non-rotating frame) until after a discussion of the interpretation of the Kerr field, and the other until Section 4.2.

---

[4] Our signature is $(-+++)$, so that $l_i n^i = -1$, $m_i \bar{m}^i = +1$.

2.2. The Interpretation of the Kerr Field

A first and fundamental property of the Kerr spacetime may be deduced immediately from Eq. (2.1.1): with the conformal factor $\Omega := r^{-1}$, the spacetime described by the metric (2.1.1) satisfies all of the conditions of Penrose's definition [31, 32, 33] of asymptotic flatness. We can therefore assert that Kerr's spacetime describes the field of an isolated body.

By transforming to a system of advanced null coordinates v, r, θ, φ' defined by

$$\mathrm{d}v := \mathrm{d}t + [(r^2 + a^2)/\Delta] \, \mathrm{d}r$$

$$\mathrm{d}\varphi' := \mathrm{d}\varphi + (a/\Delta) \, \mathrm{d}r$$

with r and θ unchanged, one extends the spacetime region covered by the BL coordinates to include the hypersurfaces on which $\Delta = r^2 + a^2 - 2Mr = 0$. In fact, in this new coordinate system the coordinate ranges are $-\infty < v < \infty$, $-\infty < r < \infty$, $0 \leq \theta \leq \pi$, $0 \leq \varphi \leq 2\pi$ provided we do not have simultaneously $r = 0$, $\theta = \pi/2$ (where Ψ_2 becomes unbounded, indicating an essential curvature singularity; see [2, 10] for details.) Also in this coordinate system, it is easy to see that $r = r_+ = M + (M^2 - a^2)^{1/2}$ is a null hypersurface, invariant under the action of the isometry group, with the topology $\mathbb{S}^2 \times \mathbb{R}^1$. Since the conformal factor used above in the demonstration of asymptotic flatness suggests that r is an appropriate radial coordinate, the future branch [59] of the null hypersurface $r = r_+$ represents the boundary of the region from which it is possible for future directed timelike or null curves to get out to asymptotic regions ($r \to \infty$ or $\Omega \to 0$), and is consequently an (absolute) event horizon in the sense of Penrose [5].

The asymptotic flatness and existence of an event horizon together are precisely the features which leads one to say that Kerr's spacetime describes a black hole. All that remains to be done in interpreting the Kerr field is to identify the parameters M and a in terms of physical quantities. This may be done in a large variety of more or less precise ways, for example by appealing to the linearized theory [9] and correspondence with Newtonian gravitation [66, 67]. It is just as easy, however, to proceed entirely within the framework of general relativity. On carrying out the conformal transformation $\mathrm{d}s \to \Omega \, \mathrm{d}s$ mentioned above, taking the limit $\Omega \to 0$, and then integrating Ψ_2 with the appropriate weighting factor (see [32]) over a spherical spacelike slice of the null hypersurface at infinity, one finds that M is just the Bondi mass [33a, 34]. Secondly, on computing the standard conserved quantity associated with the axial-symmetry[5] Killing vector ζ^i, one finds [35]

---

[5] If one were to use instead the stationary Killing vector in the expression $\xi_i(T^i_j - \frac{1}{2} T \delta^i_j)$, one would obtain the mass M as the corresponding conserved quantity.

that the angular momentum of any possible material source of the Kerr field is Ma. Thus, a is the angular momentum per unit mass. Finally, it is perhaps worth mentioning that in recent papers (see e.g. [10]) the the sign of a has been changed from that of Kerr [1, 3, 2]. Using Kerr's form of the metric [1], Boyer and Price [9] identified the angular momentum per unit mass as $-a$.

2.3. Locally Non-Rotating Frames

It has been found useful in carrying out certain calculations in the Kerr spacetime to introduce the notion of a locally non-rotating frame (LNRF) [4, 36, 42]. This may be done as follows. In Minkowski's flat spacetime there exist congruences of timelike worldlines, with respect to which one can carry out a $3+1$ spacelike-timelike decomposition of physical quantities, satisfying the conditions:

(i) they are trajectories of a 1-parameter group of isometries (i.e. the local properties of spacetime do not change along the curves),

(ii) they are hypersurface-orthogonal (so that Coriolis-type effects are absent), and

(iii) they are geodesics (i.e. freely falling, or inertial.)

In a general spacetime of course, no such congruences can be found since the three conditions require that the unit tangent of the congruence be covariantly constant. Nevertheless one can seek to find a congruence of timelike curves in the Kerr spacetime which has some properties of such a preferred flat-space congruence. It is well-known that no *constant* linear combination of the Killing vectors ξ^i and ζ^i is either hypersurface orthogonal or geodesic, so we cannot satisfy (i), (ii) and (iii) in Kerr's spacetime despite the two-parameter isometry group. Nevertheless, we can ask if there is a timelike congruence satisfying (ii), each curve of which is contained in one of the 2-dimensional group orbits. From Eq. (2.1.3) it is clear that the congruence defined by $\omega^r = \omega^\theta = \omega^\varphi = 0$, i.e. $r = \text{const}$, $\theta = \text{const}$, $d\varphi/dt = \omega$ does indeed uniquely satisfy the weakened criteria[6]. Unfortunately our congruence is not geodesic, so a frame tied to it is by no means inertial, but this is the best we can do.

Bardeen's justification [4] for singling out this particular timelike congruence is based on the following geometrical construction. Choose one particular group orbit O. From the nature of the group action (Section 2.1), O is a 2-dimensional timelike surface, with the topology

---

[6] Alternatively, the unique linear combination of the two Killing vectors which is timelike and a gradient is

$$(\Sigma \Delta)^{-1} (\mathscr{A} \xi_i + 2Mra \zeta_i) = V_i t \,.$$

of a cylinder. The curves, in O, of the chosen timelike congruence have the property that the future-directed pair of null lines on the cylinder, emanating from any point on one curve, re-intersect at a later point on the *same* curve after one revolution around the cylinder (and hence repeatedly re-intersect on this curve)[7]. The reason for calling the congruence "locally nonrotating"[8] is that, despite the fact that one of the null lines on O goes around the black hole in one direction, while the other goes around in the opposite direction, both take the same time (as measured along the curve of the congruence) to go around: the chosen timelike congruence ignores, in this sense, the rotation of the black hole.

It is possible to view this local non-rotation in yet another way. There is a well-defined sense in which the local light cones in the Kerr spacetime are "dragged over" by the rotation [24]. Bardeen's criterion enables one to select a timelike vector tangent to the group orbit at each point which "points up the centre" of the local light cone at that point in the sense that for this vector the dragging effects are cancelled.

2.4. The Ergosphere

Although the (2-dimensional) orbits of points under the action of the isometry group in the Kerr spacetime remain timelike right up to the horizon, where they become null, the Killing vector ξ^i whose norm is -1 at infinity, and which is therefore timelike near infinity, actually becomes spacelike outside the horizon [37]. The hypersurface L on which ξ^i is null is given by

$$\xi^i \xi_i = -(1 - 2 Mr/\Sigma) = 0 \,,$$

i.e.

$$r = M + (M - a^2 \cos^2 \theta)^{1/2} \,.$$

This hypersurface lies outside the horizon, except on the symmetry axis $\theta = 0, \pi$, where they touch. Moreover, L is timelike, not null (except on the axis again). This behaviour of ξ^i enables Penrose's energy extraction process [6, 8] to work. The region between the horizon and the hypersurface L has consequently been called the ergosphere [7]; the hypersurface L itself is called the ergosurface [24].

One version of the energy extraction process may be described as follows: the energy, as measured at infinity, of a test particle with 4-mo-

---

[7] Geometrically, the criterion is identical to that which enables one to select those timelike lines on a flat 2-dimensional cylindrical spacetime which do not "wind around" the cylinder.

[8] This is different from the sense in which a Fermi-Walker propagated frame is non-rotating.

mentum vector t^i is given by $E = -\xi_i t^i$. Since ξ^i is a Killing vector, E is constant along the particle's geodesic (cf. Section 3.1), and is conserved in collisions. Consider a particle with energy E_0 which enters the ergosphere from outside and then splits into two particles with energies E_1 and E_2. Since ξ^i is spacelike in the ergosphere, matters can be arranged [8] so that $E_1 > 0$ while $E_2 < 0$. If particle 1 then leaves the ergosphere (while particle 2 falls through the horizon), the relation $E_0 = E_1 + E_2$ implies that particle 1 has more energy than the original particle; energy has been extracted from the black hole. Bardeen et al. [36] have cast doubt on the astrophysical importance of this process, however, since the required initial trajectories are accessible only to very few highly energetic particles. For a different interpretation of energy extraction in terms of reversible and irreversible processes, see [64].

3. Test Particle Trajectories in the Kerr Field

3.1. First Integrals of the Equations of Motion [10, 38, 39]

The object of this section is to derive the basic relations governing the motion of freely falling test particles in the Kerr field. Since freely falling test particles follow geodesics in the spacetime, this means solving the geodesic equation. For most purposes, it is sufficient to have the components of the tangent vector to the geodesics, so we consider only the first order differential equation

$$t^i \nabla_i t^j = 0 , \tag{3.1.1}$$

where t^i is tangent to an affinely parametrized geodesic.

Now there is always one first integral of Eq. (3.1.1), namely[9] $\mu^2 := -g_{ij} t^i t^j$, where μ is the rest mass of the orbiting test particle. In Kerr's spacetime, there are in fact three additional, independent first integrals of the geodesic equation, so the problem of obtaining the components of t^i reduces to a purely algebraic one, with no integration required. Two of these constants of the motion arise from the symmetries of the Kerr spacetimes. As a consequence of Killing's equations $\nabla_{(i} \xi_{j)} = 0$, $\nabla_{(i} \zeta_{j)} = 0$ and (3.1.1), the scalars

$$E := -\xi_i t^i$$

---

[9] Throughout Section 3, and in this section only, μ denotes the rest mass of a test particle. Elsewhere, μ is as defined in 2.1.

and

$$\Phi := \zeta_i t^i$$

are constant along geodesics: we have

$$t^i \nabla_i E = -\xi_j t^i \nabla_i t^j - t^i t^j \nabla_{(i} \xi_{j)} = 0,$$

with a similar equation holding for Φ. Since ξ^i is analogous to a generator of time translations in the asymptotic region, while ζ^i generates rotations about the axis of symmetry, we identify E and Φ as the energy and angular momentum about the axis of symmetry respectively.

The third first integral arises as a consequence of the existence of a symmetric tensor K_{ij} in Kerr's spacetime [38, 39], satisfying the equation

$$\nabla_{(i} K_{jk)} = 0.$$

By analogy with the equation satisfied by a Killing vector, K_{ij} has been called a *Killing tensor*. The scalar

$$K := K_{ij} t^i t^j,$$

quadratic in the tangent vector t^i, is thus conserved along geodesics since

$$t^i \nabla_i K = 2 K_{jk} t^j t^i \nabla_i t^k + t^i t^j t^k \nabla_{(i} K_{jk)} = 0.$$

The conserved quantity K was originally found by Carter [10] as a constant of separability, in advanced null coordinates, of the Hamilton-Jacobi equation associated with (3.1.1). The physical interpretation of K is not completely clear, but as has been pointed out by Geroch [40], K reduces to the square of the total angular momentum of an orbiting test particle in the spherically symmetrical Schwarzschild spacetime.

If l^i and n^i are the repeated principal null vectors of the Kerr field, then the Killing tensor takes the form[10]

$$K_{ij} = 2\Sigma l_{(i} n_{j)} + r^2 g_{ij}. \tag{3.1.2}$$

It is useful to use Eq. (2.1.5) to reexpress K_{ij} in the form

$$K_{ij} = 2\Sigma m_{(i} \bar{m}_{j)} - a^2 \cos^2\theta \, g_{ij}. \tag{3.1.3}$$

Working in the BL coordinate system, and letting a dot denote differentiation with respect to an affine parameter τ, the tangent vector

---

[10] This is minus the corrected version of a formula given in [38] so as to agree with Carter's sign convention [10], and differs from that given in [39] because of a difference of signature; see Eq. (2.1.5).

to a general affinely parametrized geodesic may be written

$$t^i = (\dot{t}, \dot{r}, \dot{\theta}, \dot{\varphi}).$$

Using the components for the Killing vectors $\xi^i = \delta^i_t$ and $\zeta^i = \delta^i_\varphi$ from Section 2, one easily computes

$$E = (1 - 2Mr/\Sigma)\,\dot{t} + (2Mr/\Sigma)\,a\,\sin^2\theta\,\dot{\varphi} \qquad (3.1.4)$$

and

$$\Phi = -(2Mr/\Sigma)\,a\,\sin^2\theta\,\dot{t} + (\mathscr{A}/\Sigma)\,\sin^2\theta\,\dot{\varphi}. \qquad (3.1.5)$$

Similarly, from the expressions (2.1.4) for the principal null tetrad, and from (3.1.2) one finds [11]

$$K = \Delta^{-1}[(r^2 + a^2)\,E - a\Phi]^2 - \Delta^{-1}\,\Sigma^2\,\dot{r}^2 - r^2\mu^2, \qquad (3.1.6)$$

while (3.1.3) yields

$$K = \mu^2 a^2 \cos^2\theta + \Sigma^2\dot{\theta}^2 + (aE\sin\theta - \Phi/\sin\theta)^2. \qquad (3.1.7)$$

The last four equations immediately yield the desired components of t^i:

$$\dot{t} = (\Delta\Sigma)^{-1}\,(\mathscr{A}E - 2Mra\,\Phi), \qquad (3.1.8)$$

$$\Sigma^2\dot{r}^2 = [(r^2 + a^2)\,E - a\,\Phi]^2 - \Delta(\mu^2 r^2 + K), \qquad (3.1.9)$$

$$\Sigma^2\dot{\theta}^2 = K - \mu^2 a^2 \cos^2\theta - (aE\sin\theta - \Phi/\sin\theta)^2, \qquad (3.1.10)$$

$$\dot{\varphi} = \Delta^{-1}[(2Mr/\Sigma)\,aE + (1 - 2Mr/\Sigma)\,\Phi/\sin^2\theta]. \qquad (3.1.11)$$

Following Carter [10], we write

$$\Sigma^2\dot{r}^2 =: R,$$

and

$$\Sigma^2\dot{\theta}^2 =: \Theta,$$

and expand out the functions R and Θ as follows:

$$R = (E^2 - \mu^2)\,r^4 + 2M\mu^2 r^3 + 2(a^2 E^2 - \Phi^2 - a^2\mu^2 - Q)r^2$$
$$+ 2M[(aE - \Phi)^2 + Q]\,r - a^2 Q, \qquad (3.1.12)$$

$$\Theta = Q - \cos^2\theta[\Phi^2/\sin^2\theta - (E^2 - \mu^2)a^2] \qquad (3.1.13)$$

---

[11] It is useful here to use (3.1.4) and (3.1.5) to write $t_i = (-E, \Delta^{-1}E\dot{r}, \Sigma\dot{\theta}, \Phi)$, and to compute $K^{ij}t_i t_j$.

where

$$Q := K - (aE - \Phi)^2 . \tag{3.1.14}$$

Equations (3.1.8 – 11) may be integrated once more, if desired, as follows: divide (3.1.9) by (3.1.10) to obtain

$$(dr/d\theta)^2 = R(r)/\Theta(\theta) .$$

It is convenient to introduce $u := \cos\theta$ as a new independent variable, giving

$$(dr/du)^2 = R(r)/U(u) , \tag{3.1.15}$$

with

$$U(u) := Q - (Q + \Phi^2 - a^2\Gamma) u^2 - a^2\Gamma u^4 , \tag{3.1.16}$$

and

$$\Gamma := E^2 - \mu^2 . \tag{3.1.17}$$

Since R and U are both quartic polynomials, (3.1.15) can either be solved analytically in terms of elliptic functions, or numerically. A numerical solution for zero angular momentum photons ($\Phi = \mu = 0$) has been given by Floyd and Sheppee [61]. Once r is known as a function of θ, (3.1.9) or (3.1.10) can be solved for $r = r(\tau)$, $\theta = \theta(\tau)$ where τ is an affine parameter. Finally, (3.1.8) and (3.1.11) can then be solved for $t(\tau), \varphi(\tau)$. We shall not attempt this complete program here, however, but will describe in the next two sections the qualitative nature of the r- and θ-motion using only the first integrals. Motion in the equatorial plane is studied in more detail in Section 3.4.

3.2. General Features of the θ-motion

Certain aspects of the θ-motion have been discussed by Carter [10], de Felice et al. [12] and Wilkins [13]. Here we give a more complete account, starting from Eqs. (3.1.10), (3.1.13), and (3.1.16) in the form,

$$\Sigma^2 \dot{u}^2 = : f(u) = Q + Au^2 + Bu^4 , \tag{3.2.1}$$

with

$$A := -(Q + \Phi^2 - a^2\Gamma), \quad B := -a^2\Gamma. \tag{3.2.2}$$

Recall that $Q = K - (aE - \Phi)^2$, $\Gamma = E^2 - \mu^2$ and $u = \cos\theta$. It is a trivial matter to evaluate [12]

$$f(0) = Q, \quad f(1) = -\Phi^2, \quad f'(0) = 0, \quad f''(0) = 2A, \quad f'(1) = 2(2B + A).$$

Now motion is only possible for $f(u) \geq 0$, so the particle can reach the axis $u^2 = 1$ if and only if $\Phi = 0$.

If $Q > 0$, we see that $f(0) > 0$, so $f(u)$ has a zero u_0 in the range $0 < u \leq 1$. $f(u)$ can have at most one stationary point in this range. For an orbit at constant θ, say $u = u_c$, it is necessary and sufficient that $f(u)$ have a repeated zero at $u = u_c$. This can only occur at $u_c = u_0 = 1$, and the stationary point is a minimum of f. The condition for a repeated root is $Q + A + B = 0$, $A + 2B = 0$. Thus if $\Phi = 0$ and $Q = -a^2\Gamma > 0$, a particle may remain on the axis $\theta = 0$, but is unstable to small θ-perturbations. This case is illustrated in Fig. 1b. (Note that for photons $\mu^2 = 0$ and $\Gamma > 0$ so that this case cannot occur.) The general motion for $Q > 0$ is shown in Fig. 1a. The particle moves in an oscillatory fashion, repeatedly crossing the equatorial plane, with θ lying in the range $\theta_0 \leq \theta \leq \pi - \theta_0$, where

$$\cos\theta_0 := u_0, \tag{3.2.3}$$

u_0 being the unique zero of (3.2.1) in the range $0 < u \leq 1$.

If $Q = 0$, there is a trivial case with $\Gamma = \Phi = 0$. Then $\dot{u} = 0$ and θ may take any constant value. If $A < 0$ then $f(u)$ has a maximum at $u = 0$, and is otherwise negative definite. The only motion possible is $u = 0$, $\theta = \pi/2$, and is stable against small θ-perturbations. This case is illustrated in Fig. 1c. If $A > 0$ there is necessarily a maximum of f for some u_m, $0 < u_m < 1$, and $f(u_m) > 0$. In this case there is a zero u_0 of f with $0 < u_m < u_0 \leq 1$ (Fig. 1d). There is a possible motion with $u = 0$, $\theta = \pi/2$. This is unstable against small θ-perturbations, however, and the general motion is oscillatory with $\theta_0 \leq \theta \leq \pi - \theta_0$, being defined by (3.2.3). Thus we see that the condition for motion in the equatorial plane, which is stable against small θ-perturbations is

$$Q = 0, \quad \Phi^2 > a^2\Gamma. \tag{3.2.4}$$

If $Q < 0$, then $f(0) < 0$ and $f(1) \leq 0$. The only case of interest is shown in Fig. 1e. Here $A > 0$, and there is a maximum of f at u_m with $f(u_m) \geq 0$. This maximum can only occur if $4BQ \geq A^2$ (which requires $\Gamma > 0$), or

$$(Q + \Phi^2 - a^2\Gamma)^2 + 4a^2\Gamma Q \leq 0. \tag{3.2.5}$$

---

[12] A prime will always (except in Section 5) denote differentiation with respect to the argument of the function.

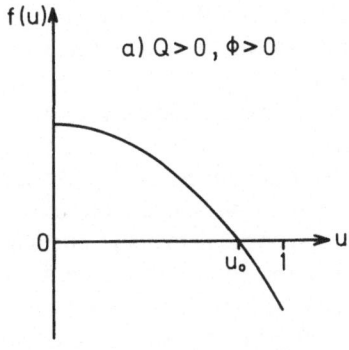

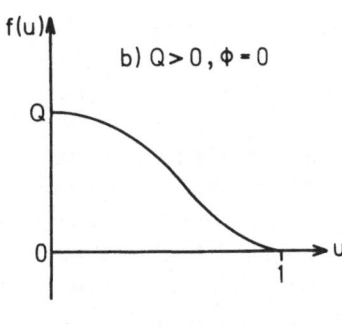

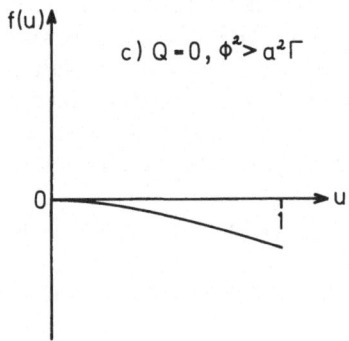

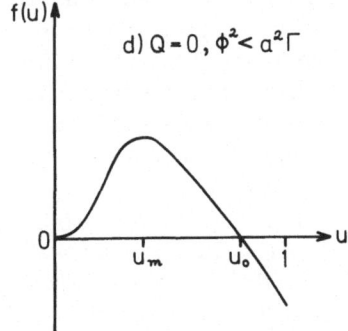

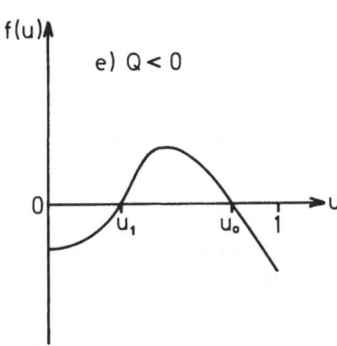

Fig. 1. *The θ-Motion.* $u = \cos\theta$, and $f(u)$ is defined by Eq. (3.2.1). Since $\dot{u}^2 \propto f$, motion is only possible for $f > 0$. The particle can reach the axis ($u = 1$) only if $\Phi = 0$. If in addition $Q > 0$ the particle can remain there in unstable equilibrium. For $Q > 0$, the general motion is oscillatory with $\theta_0 \leqq \theta \leqq \pi - \theta_0$, ($\theta_0 = \cos^{-1} u_0$). If $Q = 0$, there is an equilibrium position at $\theta = \pi/2$, $u = 0$, which is stable if $\Phi^2 > a^2 \Gamma$. If $Q < 0$ the particle oscillates in a range $0 \leqq \theta_1 \leqq \theta \leqq \theta_0 < \pi/2$ which does not include the equatorial plane. If $\theta_1 = \theta_0$, there is a position of stable equilibrium

A necessary condition for (3.2.5) to be satisfied is

$$\Phi^2 + a^2 \Gamma > 0, \quad \Gamma > 0. \tag{3.2.6}$$

If equality occurs in (3.2.5), there is a repeated root at $u_m = u_0$. The motion is thus given by $\theta = \theta_m = \cos^{-1} u_m$, which is stable against small θ-perturbations. In general, the motion is oscillatory with $0 < u_1 \leqq u \leqq u_0 \leqq 1$ where u_0 and u_1 are the two positive zeros of f. Notice that in this case the particle never crosses the equatorial plane.

These results are of particular interest if one considers accretion processes [26] or gravitational synchrotron radiation (Section 4.5). One normally studies accretion only of particles moving in the equatorial plane, because this is the easiest case. For this to be realistic one must show a) that the motion is stable, and b) that typical realistic particle motion will tend to evolve towards motion in the equatorial plane. It is plausible that a particle moving in an orbit around a Schwarzschild black hole could reduce the quantities E, Φ, and K toward zero by emitting gravitational radiation, since K represents the square of the total angular momentum. Such a straightforward interpretation for K does not appear to exist for a Kerr ($a \neq 0$) black hole. Even if one were able to show that $K \to 0$ for a particle emitting gravitational radiation while orbiting a Kerr black hole, this would not be sufficient to demonstrate that such orbits eventually end up in the equatorial plane. To be sure that they do end up in the equatorial plane one would need to show that

1) for the $Q > 0$ orbits the effect of gravitational radiation is to cause $Q \to 0$ keeping $\Phi^2 > a^2 \Gamma$;

2) For the $Q < 0$ orbits the effect of gravitational radiation is to cause $Q \to 0$ keeping $\Phi^2 + a^2 \Gamma > 0$.

3.3. General Features of the *r*-motion

Equations (3.1.9) and (3.1.12) can be written

$$\Sigma^2 \dot{r}^2 = R(r) = \Gamma r^4 + 2M\mu^2 r^3 + (a^2 E^2 - \Phi^2 - a^2 \mu^2 - Q)r^2$$
$$+ 2M[(aE - \Phi)^2 + Q] r - a^2 Q. \tag{3.3.1}$$

We note that on the horizon $r = r_+$, $\Delta = 0$. Therefore Eq. (3.1.9) implies $R(r_+) \geqq 0$. We need also to note that $R(0) = -a^2 Q$, $R'(0) = 2M[(aE - \Phi)^2 + Q]$, $R(r) \to \Gamma r^4$ as $r \to \infty$, and

$$\partial R/\partial \Phi = -4aEMr - 2\Phi r(r - 2M). \tag{3.3.2}$$

We shall distinguish 4 cases according as $\Gamma \gtrless 0$, $Q \gtrless 0$. The limiting case $\Gamma = 0$, $E = 1$ is easily found by inspection, and the case $Q = 0$ will be treated in more detail in Section (3.4). Carter [10] and Floyd and Sheppee [61] have discussed some features of the general *r*-motion, and Wilkins [13] has discussed the case $\Gamma > 0$. (References to work concerned purely with orbits in the equatorial plane will be given in the next section.)

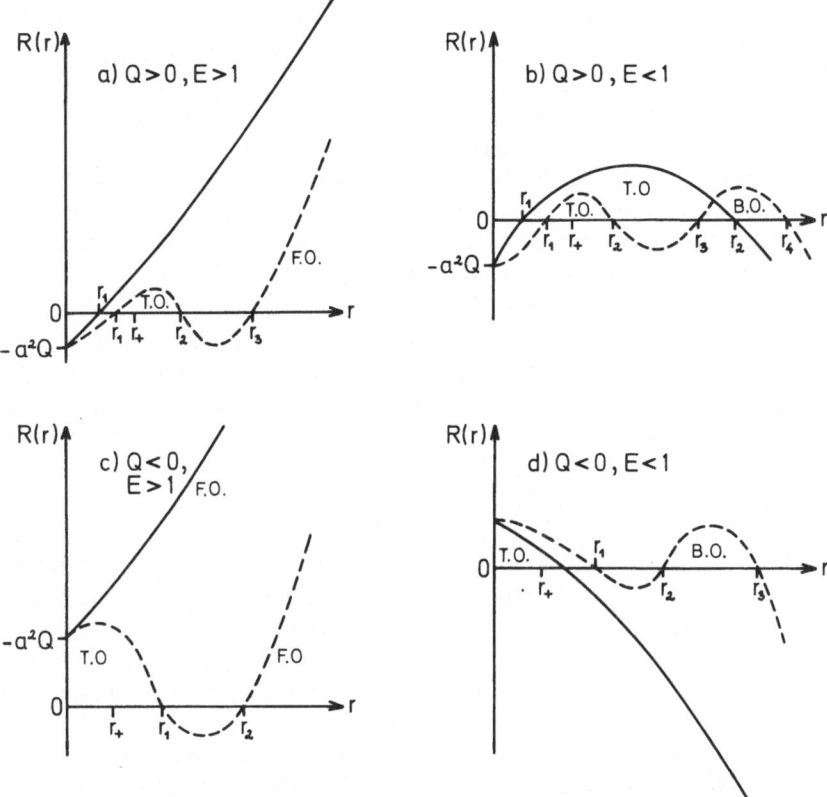

Fig. 2. *The r-Motion.* $R(r)$ is defined by Eq. (3.1.12). Since $\dot{r}^2 \propto R$, motion is only possible for $R > 0$. Particles which are constrained to move in a range of r which includes the horizon r_+ move on *trapped orbits* (T.O.), and must eventually disappear inside the horizon. Those constrained to a finite range of r outside r_+ move on *bound orbits* (B.O.), while those moving in a semi-infinite range of r move along *free orbits* (F.O.). The solid lines indicate $R(r)$ for large negative values of Φ (retrograde orbits), while the dotted lines refer to large positive values of Φ (prograde orbits)

Case a: $Q > 0$, $E > 1$. If E is sufficiently large, all the coefficients in (3.3.1) except that of r^0 are non-negative. A sketch of $R(r)$ is shown in Fig. 2a as a solid line. Since $f(0) < 0$, $f(r_+) \geqq 0$, there must be a zero of R, $r = r_1$, in or on the horizon. Suppose Φ is now increased to some large positive value, keeping E and Q fixed. It is evident from (3.3.2) that the shape of the curve will alter to the dotted curve in Fig. 2a. There will now be a second zero near $r = r_2 \simeq 2M$, and a third zero at $r_3 > 2M$. There are no other possible positions for the zeros in this case. Now consider a particle moving inwards from radial infinity in the first case (solid line). Since $\dot{r}^2 > 0$ the particle will move inwards and cross the horizon until r_1 is reached. Here $\dot{r} = 0$, and the particle reverses its motion. However it cannot recross the horizon in finite coordinate or proper time (or indeed at all); it has been *trapped*. A particle outside the horizon with $\dot{r} > 0$ will escape to infinity; it is *free*. In the second case, (dotted line), any particle incident from infinity will be reflected back to infinity at $r = r_3$. No particle can have r in the range $r_2 < r < r_3$. Any particle with $r < r_2$ is trapped. For, even if it is moving outwards, it will be reflected inwards at $r = r_2$ and must ultimately cross the horizon. There is a special case when $r_2 = r_3$. Then the particle moves at constant r, a *spherical orbit*. However such an orbit is unstable to r-perturbations.

Case b: $Q > 0$, $E < 1$. Since $f(0) < 0$, $f(1) \geqq 0$, $f(r) \to \infty$ as $r \to \infty$, there must be at least two real zeros of f (possibly coincident on the horizon). The case in which there are only two real zeros is shown as a solid curve in Fig. 2b. Since $r_1 \leqq r_+ \leqq r_2$ all particles are trapped. In the limiting case $r_1 = r_2$ there is an equilibrium position with the particle orbiting on the horizon. At first sight one might expect this equilibrium to be stable. However if a small r-perturbation is made the particle must eventually disappear within the horizon. It is clear that by changing Φ, keeping E and Q constant, the situation depicted by the dotted line in Fig. 2b can also occur. As before, particles in the range $r_1 \leqq r \leqq r_2$ are trapped. However there are two more zeros r_3 and r_4 of f and any particle in the range $r_3 \leqq r \leqq r_4$ remains oscillating in that range: it is *bound*. Clearly there is a limiting case when $r_3 = r_4$, and this gives a *stable spherical orbit*. Using (3.3.2) it is easy to see that the negative Φ spherical orbits are at a larger r than the positive Φ ones.

Case c: $Q < 0$, $E > 1$. By inspection of diagram 2c, it will be seen that this case is very similar to case a, so it is only necessary to point out that all trapped particles eventually reach $r = 0$. There is no possibility of four real positive zeros in this case, because their sum would have to be $-2M\mu/\Gamma \leqq 0$.

Case d: $Q < 0$, $E < 1$. This case is very similar to $2d$, except that now, all trapped particles can reach $r = 0$.

These results may be summarized as follows. If $E > 1$ any particle which is sufficiently far from the black hole, and is moving away from it, will escape to infinity. An incoming particle from infinity will be repelled if it has a sufficiently large positive angular momentum component Φ. It is also possible for particles with large positive Φ to exist closer to the black hole. However they can never escape to infinity and will always be trapped by the horizon. If the angular momentum Φ is not sufficiently large, then all incoming particles are trapped. However no trapped particle reaches $r = 0$ unless $Q \leqq 0$. If $E < 1$ no particle can reach infinity; any particle is either trapped or bounded. If the particle is bounded, then stable spherical orbits can exist. Those for $\Phi > 0$ are closer to the black hole than those for $\Phi < 0$, and there is a limiting case with the former at the horizon (but no longer stable!).

3.4. Motion in the Equatorial Plane

The case $Q = 0$, $\theta = \pi/2$ has been widely studied [2, 7, 10, 11, 13, 26, 36], although as we have seen in Section (3.2) only those particles with $\Phi^2 > a^2 \Gamma$ are stable to small θ-perturbations. In this section we shall assume for convenience that $\mu = 1$, that is, we shall assume the particles have non-zero restmass. The photon case can always be determined by the limit $E \to \infty$. When $Q = 0$, we may divide Eq. (3.3.1) by r to obtain

$$r^{-1} \Sigma^2 \dot{r}^2 = g(r) = \Gamma r^3 + 2Mr^2 + (a^2 \Gamma - \Phi^2) r + 2M(aE - \Phi)^2 . \tag{3.4.1}$$

We can quickly dispose of the orbits which are unstable to θ-perturbations. For, if $\Phi^2 < a^2 \Gamma$, then all coefficients in (3.4.1) are non-negative. g therefore has no real zeros. Incoming particles proceed inwards to $r = 0$, while particles outside the horizon which are moving outwards will escape to infinity. Henceforth we consider only particles which are stable against θ-perturbations.

First consider the case $\Gamma > 0$. Since the sum of the zeros of g is negative and $g(0) \geq 0$, $g(r) \to +\infty$ as $r \to \infty$, g may have 0 or 2 positive zeros. If there are no zeros we have the case shown by the solid in Fig. 2c. If there are 2 zeros we have the case shown by the dotted line. If $\Gamma < 0$ then $g(r) \to -\infty$ as $r \to \infty$ and g may have 1 or 3 positive zeros. These possibilities are shown in Fig. 2d. In both cases the qualitative description exactly mirrors that for $Q < 0$ given in the previous section.

However when $Q = 0$ it is possible without too much effort to give much more precise information about the orbits. It is a trivial piece of algebra to show that (3.4.1) can have a repeated root if and only if

$$[\Phi^2 - a^2 \Gamma + 9\Gamma(aE - \Phi)^2]^2 - (\Phi^2 - a^2 \Gamma) [\Phi^2 - a^2 \Gamma + 9\Gamma(aE - \Phi)^2]$$
$$\times [3\Gamma(\Phi^2 - a^2 \Gamma) + 4M^2] \tag{3.4.2}$$
$$+ 3(aE - \Phi)^2 [3\Gamma(\Phi^2 - a^2 \Gamma) + 4M^2]^2 = 0 ,$$

and the value of the repeated zero is given by

$$[3\Gamma(\Phi^2 - a^2 \Gamma) + 4M^2] r = M[\Phi^2 - a^2 \Gamma + 9\Gamma(aE - \Phi)^2] . \tag{3.4.3}$$

We first find the photon orbits. Letting $y = \Phi/E$, and taking the limit of (3.4.2) and (3.4.3) as E and $\Phi^2 \to \infty$ with y finite, we obtain

$$y^3 + 3ay^2 + 3(a^2 - 9M^2) y + a(a^2 + 27M^2) = 0 , \tag{3.4.4}$$

$$r_y/M = 3(y - a)/(y + a) . \tag{3.4.5}$$

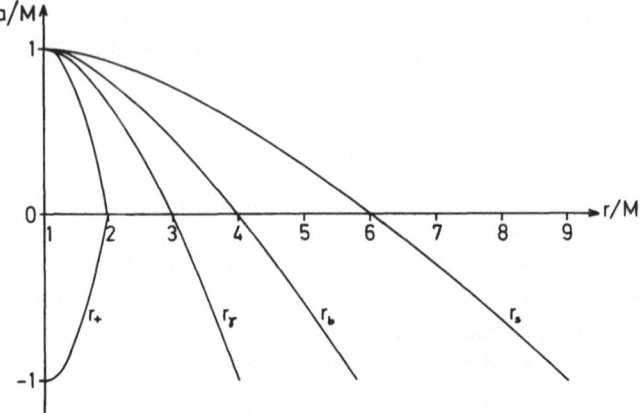

Fig. 3. *Equatorial Orbits for the Kerr Spacetime.* Shown here are the radii of some of the circular orbits in the Kerr spacetime as a function of the Kerr parameter a. Positive values of a refer to prograde orbits, negative values to retrograde orbits. r_+ gives the position of the horizon, r_γ the circular photon orbit, r_b the radius of the marginally bound ($E = 1$) orbit, and r_s the radius of the smallest stable orbit. The apparent merging at $r = a = M$ is a coordinate effect (see Section 3.4 and Fig. 5)

Eliminating y, we have

$$r_\gamma^2 - 3Mr_\gamma \pm 2a(Mr_\gamma)^{1/2} = 0\,, \tag{3.4.6}$$

where the upper sign corresponds to prograde ($\Phi > 0$) orbits, and the lower sign to retrograde ($\Phi < 0$) orbits. For $a = 0$ we obtain $r_\gamma = 3M$ and $\Phi = \pm 3\sqrt{3}ME$, the usual Schwarzschild result. For $a = 1$ we obtain $\Phi = 2ME$ and $r_\gamma = M$ for prograde orbits, $\Phi = -7ME$ and $r_\gamma = 4M$ for retrograde orbits. r_γ is shown as a function of a in Fig. 3. Notice that since $\Gamma > 0$, these circular orbits correspond to a repeated root in Fig. 2c and so are unstable to r-perturbations.

Another simple case which is easily discussed is that when $E = 1$, in which a particle can just escape to infinity. For such a "marginally bound" particle to have a circular orbit at $r = r_b$, (3.4.3) requires

$$r_b = \Phi^2/(4M)\,, \tag{3.4.7}$$

while (3.4.2) implies

$$\Phi^4 = 16\,M^2(\Phi - a)^2\,. \tag{3.4.8}$$

Eliminating Φ gives

$$r_b^2 - 4Mr_b \pm 4a(Mr_b)^{1/2} - a^2 = 0\,. \tag{3.4.9}$$

In the Schwarzschild case we obtain $\Phi = \pm 4M$ and $r_b = 4M$. In the extreme $a = M$ Kerr case we have $r_b = M$ and $\Phi = 2M$ for prograde orbits, $r_b = (3 + 2\sqrt{2})M$ and $\Phi = -2M(1 + \sqrt{2})$ for retrograde orbits. r_b is shown as a function of a in Fig. 3. These circular orbits are of course unstable to small r-perturbations.

For a stable circular orbit it is necessary and sufficient that $g(r) = g'(r) = 0$ while $g''(r) < 0$. Since $g''(r) = 6\Gamma r + 4M$, such orbits are only possible for bound particles with $E < 1$. The division between stable and unstable circular orbits occurs when $g''(r) = 0$, say $r = r_s = -2M/3\Gamma$. This requires

$$\Phi^2 = a^2\Gamma - 4M/3\Gamma, \tag{3.4.10}$$

$$(aE - \Phi)^2 = 4M^2/(27\Gamma^2). \tag{3.4.11}$$

Solving for r_s gives

$$r_s^2 - 6Mr_s \pm 8a(Mr_s)^{1/2} - 3a^2 = 0. \tag{3.4.12}$$

In the Schwarzschild case $r_s = 6M$, $E = 2\sqrt{2}/3$ and $\Phi = \pm 2\sqrt{3}\,M$. In the extreme $a = M$ Kerr case, $r_s = M$, $E = 1/\sqrt{3}$ and $\Phi = 2M/\sqrt{3}$ for prograde orbits, $r_s = 9M$, $E = 5/3\sqrt{3}$ and $\Phi = -22/(3\sqrt{3})\,M$ for retrograde orbits. r_s is shown as a function of a in Fig. 3. Eqs. (3.4.6), (3.4.9) and (3.4.12) have been previously obtained by computerised algebra techniques [36].

Considerably more information is contained in Figs. 4 and 5, which apply to the Schwarzschild $a = 0$ and extreme $a = M$ Kerr cases. To understand these one should realise that since $Q = 0$ and $\mu = 1$ the motion is completely specified by giving the other 2 first integrals E and Φ, plus initial data. Thus the $E - \Phi$ space is a phase space for the particles. Specifiying E and Φ determines the motion of the particle up to a possible ambiguity in direction, and an ambiguity in position if two disjoint ranges of r are permitted. In Fig. 4 the solid lines represent the circular orbits, labelled according to stability. The two cusps occur at the point of marginal stability $r = r_s$. The point where $r = r_b(E = 1)$ is also marked. The circular photon orbits can be found by following the solid lines out to $E \rightarrow \infty$. These solid lines plus the line $E = 1$ divide the phase space into disjoint regions in which the shape of the $\dot{r}^2(r)$ curve is qualitatively different. Small sketches are provided to assist the reader in determining the nature of the orbits, and it is easy to identify the regions for free, bounded and trapped orbits. Figure 5 shows the phase space for the extreme Kerr $a = M$ case. Now the regions are not qualitatively different. The diagram can be obtained from Fig. 4 by a clockwise rotation plus some distortion. Thus to get from a point in Fig. 4 to the

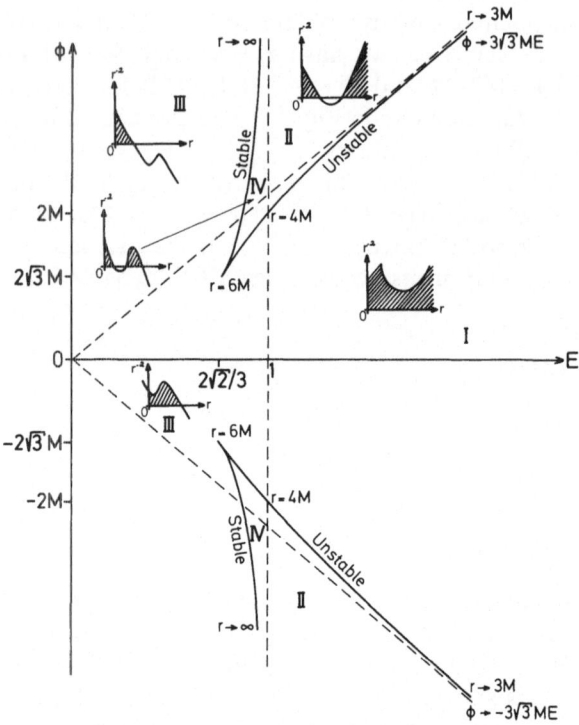

Fig. 4. *Timelike Geodesic Orbits of the Schwarzschild Spacetime.* Each timelike geodesic is specified almost uniquely (up to a possible ambiguity in direction) by the constants of motion E, Φ. The solid lines represent the circular orbits $r =$ constant. They and the line $E = 1$ divide the phase space into regions labelled I-IV. For each region a sketch of $\dot{r}^2(r)$ is provided, the shaded areas of which indicate the possible motion. In region I particles may travel in and out without hindrance. In region II, particles coming in from large r are reflected outwards (a *free* orbit), outgoing particles at small r are reflected inwards (a *trapped* orbit). In region III, all orbits are trapped. In region IV, orbits are either trapped or *bounded* (oscillating between two finite nonzero values of r)

corresponding point in Fig. 5 it is necessary to decrease Φ (another frame dragging effect). Note that Fig. 5 is simple because all information concerning orbits lying totally within the horizon (for example those with $r < 0$) have been omitted.

The following general points should be noted. A particle cannot be in a circular orbit unless E is greater than a certain minimum value, which in the Kerr $(a \neq 0)$ case is smaller for prograde than for retrograde orbits. A particle cannot reach $r = 0$ by moving inwards from infinity unless it lies in region I, so that its momenta must lie in some range $-\Phi_1(E) < \Phi < \Phi_2(E)$, where E is its energy. In Schwarzschild, $\Phi_1 = \Phi_2$, but

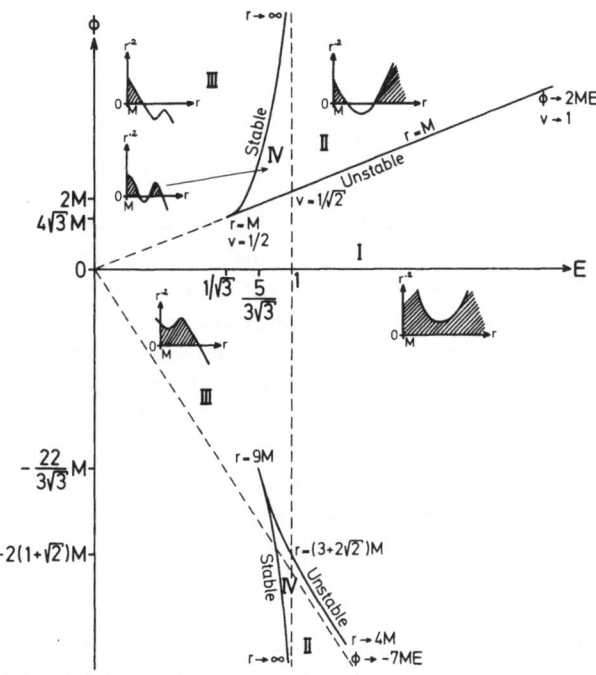

Fig. 5. *Timelike Geodesic Orbits of the Kerr $a = M$ Spacetime.* Each timelike geodesic is specified almost uniquely (up to a possible ambiguity in direction) by the constants of motion E, Φ. The solid lines represent the circular orbits $r = $ constant. They and the line $E = 1$ divide the phase space into regions labelled I-IV. For each region a sketch of $\dot{r}^2(r)$ is provided, the shaded areas of which indicate the possible motion. In region I particles may travel in and out without hindrance. In region II, particles coming in from large r are reflected outwards (a *free* orbit), outgoing particles at small r are reflected inwards (a *trapped* orbit). In region III, all orbits are trapped. In region IV, orbits are either trapped or *bounded* (oscillating between two finite nonzero values of r). Also shown are values of v, the azimuthal velocity defined by Eq. (4.2.3), for some circular orbits. Note that no details of orbits lying inside the horizon are given in this diagram

in general $\Phi_2 < \Phi_1$ (dragging of frames again). Finally one should note what happens at $r = M$ in the extreme Kerr $a = M$ case. From Eqs. (3.4.6), (3.4.9) and (3.4.12) or Fig. 3 one might suppose that the photon, marginal bound and smallest stable orbits coincide at $r = M$. However these orbits correspond to points on Fig. 5 which are distinct. The paradox arises because in Boyer-Lindquist coordinates the metric (2.1.1) has $g_{rr} = \Sigma/\Delta$, and Δ vanishes on the horizon. In fact the orbits have non-zero proper (spacelike) separation, even though the coordinate separation vanishes.

4. Tidal Forces

4.1. The Basic Theorem

The behaviour of test bodies moving at relativistic velocities near black
holes is of great interest. The work of Pirani [41] suggests that for such
bodies, the tidal accelerations can be magnified by a kinematical factor so
that a fast moving body could be torn apart. It is clearly important to
determine under what circumstances this kinematic amplification takes
place. We first compute the tidal accelerations experienced by a test
particle moving along a geodesic in an arbitrary spacetime. This formula
appears rather complicated, but drastic simplifications occur for algebraically special vacuum spacetimes, such as those of Kerr and Schwarzschild. From the general formula one can deduce that, if the 4-velocity
of the particle is "close", in a sense to be defined, to a repeated principal
null vector of the spacetime, then no kinematical magnification takes
place. The applications of this result will occupy the rest of the chapter.

The Riemann tensor is defined here by

$$2 V_{[i} V_{j]} u_k = R_{ijkl} u^l \tag{4.1.1}$$

for any vector u^i where V_i denotes the usual covariant derivative. Now
suppose that u^i is the unit tangent vector to a timelike geodesic, so that if
$D := u^i V_i =: V_u$,

$$D u^i = 0. \tag{4.1.2}$$

Let η^i be any unit vector orthogonal to, and commuting with u^i. The
Jacobi or geodesic deviation equation is then

$$D^2 \eta^i = R^i_{jkl} u^j \eta^k u^l. \tag{4.1.3}$$

Here $\alpha\eta^i$, $\alpha \ll 1$, can be thought of as a vector connecting a pair of infi-
nitesimally neighbouring geodesics; $\alpha D^2 \eta^i$ is then their relative acceleration. Let η'^i be any other unit vector orthogonal to u^i. Then the component of the tidal acceleration along η'^i is

$$T = \eta'_i D^2 \eta^i = R_{ijkl} \eta'^i u^j \eta^k u^l. \tag{4.1.4}$$

We now introduce an NP null tetrad l^i, n^i, m^i, $\bar{m}^i$, and the usual curvature
quantities Ψ's, Φ's, and Λ defined by Newman and Penrose [29]. If we
obtain the tetrad components of u^i, η^i, η'^i as follows:

$$u^i = A l^i + B n^i + \bar{C} m^i + C \bar{m}^i \tag{4.1.5a}$$

$$\eta^i = X l^i + Y n^i + \bar{Z} m^i + Z \bar{m}^i \tag{4.1.5b}$$

$$\eta'^i = X' l^i + Y' n^i + \bar{Z}' m^i + Z' \bar{m}^i, \tag{4.1.5c}$$

then a straightforward computation gives

$$T = \Psi_0[(\bar{C}X - A\bar{Z})(\bar{C}X' - A\bar{Z}')]$$
$$+ \Psi_1[(\bar{C}X - A\bar{Z})(BX' - AY' + \bar{C}Z' - C\bar{Z}')$$
$$+ (\bar{C}X' - A\bar{Z}')(BX - AY + \bar{C}Z - C\bar{Z})]$$
$$+ \Psi_2[(BZ - CY)(\bar{C}X' - A\bar{Z}')$$
$$+ (BX - AY + \bar{C}Z - C\bar{Z})(BX' - AY' + \bar{C}Z' - C\bar{Z}')$$
$$+ (BZ' - CY')(\bar{C}X - A\bar{Z})]$$
$$+ \Psi_3[(BZ - CY)(BX' - AY' + \bar{C}Z' - C\bar{Z}')$$
$$+ (BZ' - CY')(BX - AY + \bar{C}Z - C\bar{Z})]$$
$$+ \Psi_4[(BZ - CY)(BZ' - CY')]$$
$$+ 2\Lambda(AB - C\bar{C})(X'Y + XY' - Z'\bar{Z} - Z\bar{Z}') \qquad (4.1.6)$$
$$+ \Phi_{00}(CX' - AZ')(\bar{C}X - A\bar{Z})$$
$$+ [\Phi_{01}(\bar{C}X - A\bar{Z}) + \Phi_{21}(BZ - CY)](BX' - \bar{C}Z' + C\bar{Z}' - AY')$$
$$+ [\Phi_{10}(CX' - AZ') + \Phi_{12}(B\bar{Z}' - \bar{C}Y')](BX - C\bar{Z} + \bar{C}Z - AY)$$
$$+ \Phi_{02}(B\bar{Z}' - \bar{C}Y')(\bar{C}X - A\bar{Z})$$
$$+ \Phi_{11}(BX - C\bar{Z} + \bar{C}Z - AY)(BX' - \bar{C}Z' + C\bar{Z}' - AY')$$
$$+ \Phi_{20}(BZ - CY)(CX' - AZ') + \Phi_{22}(B\bar{Z}' - \bar{C}Y')(BZ - CY)$$

+ complex conjugate .

This formula appears rather complicated. Suppose, however, that l^i is tangent to a repeated principal null direction (RPND) in an algebraically special vacuum spacetime. Then the tetrad can be chosen to make the following specialisations:

Type II, $\{211\}$: $\Psi_0 = \Psi_1 = \Psi_4 = 0$

D, $\{22\}$: $\Psi_0 = \Psi_1 = \Psi_3 = \Psi_4 = 0$

III, $\{31\}$: $\Psi_0 = \Psi_1 = \Psi_2 = \Psi_4 = 0$ $\qquad (4.1.7)$

N, $\{4\}$: $\Psi_0 = \Psi_1 = \Psi_2 = \Psi_3 = 0$.

Of course all of the Φ's and Λ are zero in the vacuum case too. We can now deduce the following

Theorem:[13] Consider an algebraically special vacuum spacetime with repeated principal null direction l^i and a sequence of timelike

---

[13] A number of inaccuracies crept into the account [14] which we wish to correct here. The statement of the Theorem should read, "Let $(\mathcal{M}, g_{ab})$ be a vacuum spacetime ...", and, "$T = O(1)$ for type II ($\{211\}$)". The expression $T = O(m)$ in the first paragraph of Section 3 should read $T = O(m^{-2})$. Finally the general formula should include the square brackets as in (4.1.6) above. See also the remark following the proof of the theorem.

geodesics through some point p whose unit tangent vectors $u^i(\varepsilon)$ at p differ from each other by a sequence of boosts in a timelike plane containing l^i. Suppose that the limit of the directions defined by this sequence $u^i(\varepsilon)$ of vectors, as $\varepsilon \to 0$, is the direction [14] of l^i. If $\eta^i(\varepsilon)$ and $\eta'^i(\varepsilon)$ are unit spacelike vectors at p, orthogonal to $u^i(\varepsilon)$, then a typical component

$$T(\varepsilon) = R_{ijkl}\eta'^i(\varepsilon)\, u^j(\varepsilon)\, \eta^k(\varepsilon)\, u^l(\varepsilon)$$

of the tidal acceleration of a neighbouring geodesic has the behaviour

$$T(\varepsilon) = O(1)$$

as $\varepsilon \to 0$.

In other words, no kinematical magnification of tidal accelerations occurs for timelike geodesics preferentially oriented with respect to the curvature in algebraically special spacetimes.

Proof: For convenience all timelike or null vectors are taken to be future-directed. Let n^i be the other null vector (besides l^i) in the timelike plane containing the sequence $u^i(\varepsilon)$, and let m^i be a complex null vector whose real and imaginary parts lie in the spacelike plane orthogonal to l^i and n^i. Suppose that l^i, n^i, m^i are normalized so as to form an NP tetrad at p. Then expand the vectors $u^i(\varepsilon)$, $\eta^i(\varepsilon)$, $\eta'^i(\varepsilon)$ so that they have components in this tetrad as follows:

$$u^i(\varepsilon) = A(\varepsilon)\, l^i + B(\varepsilon)\, n^i + \bar{C}m^i + C\bar{m}^i$$

$$\eta^i(\varepsilon) = X(\varepsilon)\, l^i + Y(\varepsilon)\, n^i + \bar{Z}(\varepsilon)\, m^i + Z(\varepsilon)\, \bar{m}^i$$

$$\eta'^i(\varepsilon) = X'(\varepsilon)\, l^i + Y'(\varepsilon)\, n^i + \bar{Z}'(\varepsilon)\, m^i + Z'(\varepsilon)\, \bar{m}^i\,.$$

The statement of the theorem then implies that

$$A = O(\varepsilon^{-1}),\quad B = O(\varepsilon),\quad C = O(1)\,. \tag{4.1.8}$$

It is in the sense of Eq. (4.1.8) that we will speak of u^i as being "close" to the null vector l^i. The orthonormality relations among the vectors u^i, η^i and η'^i, together with (4.1.8), imply that

$$X = O(\varepsilon^{-1}),\quad Y = O(\varepsilon),\quad Z = O(1)$$
$$X' = O(\varepsilon^{-1}),\quad Y' = O(\varepsilon),\quad Z' = O(1)\,. \tag{4.1.9}$$

Since we are considering a vacuum spacetime, the Φ's and Λ in Eq. (4.1.6) vanish, and since the spacetime is algebraically special with l^i tangent to a RPND, $\Psi_0 = \Psi_1 = 0$. Examination of the surviving terms in (4.1.6) using the expressions (4.1.8) and (4.1.9) completes the proof.

---

[14] The limit as $\varepsilon \to 0$ of the sequence of vectors itself does not, of course, exist. The limit of the directions, on the other hand, remains well defined. We thank B. Schmidt for this remark.

Remark: If the vector n^i above coincides with the null vector required to produce the specializations in (4.1.7), then the stronger result

$T = O(1)$ for types $\{211\}, \{22\}$

$T = O(\varepsilon)$ for type $\{31\}$

$T = O(\varepsilon^2)$ for type $\{4\}$

holds, i.e. the stronger result [14] depends on the manner in which the direction of l^i is approached.

As illustrations of the theorem, consider a Schwarzschild spacetime, (type D, $\{22\}$). Here we can take l^i to be the inward-directed radial null direction. The theorem then gives the well known result that a radially infalling body experiences tidal accelerations $T = O(M^{-2})$ as it crosses the Schwarzschild horizon at $r = 2M$, with no peculiar kinematic effects. The other interesting null orbits in the Schwarzschild spacetime are the circular null geodesics at $r = 3M$. If we consider a circular timelike geodesic at $r = 3M(1 + \varepsilon)$, we find A, B, $C = O(\varepsilon^{-1/2})$ as $\varepsilon \to 0$. Thus the theorem is not satisfied, and in fact $T = O(\varepsilon^{-1})$. This is because the orbit is not preferentially oriented with respect to a repeated principal null direction. Examples in Kerr spacetimes are treated in the next sections.

4.2. Tidal Forces in Circular Equatorial Orbits

In this section we discuss some results obtained (in a somewhat more complicated manner) by Fishbone [15, 42]. We consider a test particle moving in a circular orbit in the equatorial plane $\theta = \pi/2$ of a Kerr spacetime. Writing $\alpha = (r^2 \Delta / \mathscr{A})^{1/2}$, $\mu = (\Delta/r^2)^{1/2}$, $\beta = (\mathscr{A}/r^2)^{1/2}$, the locally non rotating frame in the equatorial plane is given by Eq. (2.1.3),

$$\omega^t = \alpha \, dt, \quad \omega^r = \mu^{-1} \, dr, \quad \omega^\theta = r \, d\theta, \quad \omega^\varphi = \beta(d\varphi - \omega \, dt), \qquad (4.2.1)$$

with the dual tetrad of vectors,

$$e_t = \alpha^{-1}(\partial/\partial t + \omega \, \partial/\partial \varphi), \quad e_r = \mu \, \partial/\partial r, \quad e_\theta = r^{-1} \, \partial/\partial \theta, \quad e_\varphi = \beta^{-1} \, \partial/\partial \varphi. \qquad (4.2.2)$$

Suppose the test body is moving with angular velocity[15] $\Omega(r) := d\varphi/dt$. Its azimuthal velocity is

$$v := \omega^\varphi / \omega^t = (\beta/\alpha)(\Omega - \omega), \qquad (4.2.3)$$

and its 4-velocity is

$$u := \gamma(e_t + v e_\varphi), \qquad (4.2.4)$$

---

[15] This Ω has nothing to do with the Ω of Section 2.2.

where $\gamma := (1 - v^2)^{-1/2}$. It is convenient to introduce a tetrad $e_\tau, e_r, e_\theta, e_\psi$ which comoves with the body. We clearly have

$$e_\tau = u = \gamma(e_t + v e_\varphi)$$

and so

$$e_\psi = \gamma(v e_t + e_\varphi), \tag{4.2.5}$$

with e_r and e_θ unchanged. The dual basis has

$$\omega^\tau = \gamma(\omega^t - v \omega^\varphi), \quad \omega^\psi = \gamma(-v \omega^t + \omega^\varphi), \quad \omega^r, \omega^\theta \quad \text{unchanged}, \tag{4.2.6}$$

which can be inverted to give

$$\omega^t = \gamma(\omega^\tau + v \omega^\psi), \quad \omega^\varphi = \gamma(v \omega^\tau + \omega^\psi). \tag{4.2.7}$$

Let η^i be a spacelike connecting vector orthogonal to u^i, and let

$$D := u^i V_i = V_u = V_{e_\tau} = : V_\tau$$

be the directional derivative along the body's worldline. The geodesic deviation equation is

$$D^2 \eta^i = R^i_{jkl} u^j \eta^k u^l. \tag{4.2.8}$$

Clearly η has to be a linear combination of e_r, e_θ, e_ψ, so writing $\eta = : x^\alpha e_\alpha$, $\alpha := r, \theta, \psi$, we have

$$D\eta = (Dx^\alpha) e_\alpha + x^\alpha De_\alpha.$$

Now $De_\alpha = V_\tau e_\alpha = : \Gamma^a_{\alpha\tau} e_a$, where $\Gamma_{abc} = \Gamma_{[ab]c} = e_a V_c e_b$ is a tetrad rotation coefficient. Since u is geodesic, $V_\tau e_\tau = 0$, and

$$\Gamma^a_{\tau\tau} = 0 = \Gamma^\tau_{a\tau}.$$

Therefore we can write

$$V_\tau e_\alpha = \Gamma^\beta_{\alpha\tau} e_\beta, \tag{4.2.9}$$

a spacelike infinitesimal rotation. To calculate the rotation coefficients Γ^a_{bc} it is usually quickest to use the dual of the definition given here. For an orthonormal frame one calculates the 1-forms ω_{ab} defined by

$$d\omega^a = -\omega^a{}_b \wedge \omega^b, \quad \omega_{ab} = \omega_{[ab]},$$

and expands

$$\omega^a{}_b = \Gamma^a_{bc} \omega^c.$$

Since the rotation coefficients have been calculated in the LNRF [36] one can simply make the required transformation between frames. This is complicated, however, since the Γ^a_{bc} are not tensors. However, for (4.2.9) we only need $\Gamma^\alpha_{\beta\tau}$ so it suffices to calculate ω^r_θ, ω^r_ψ, ω^θ_ψ, which we can do by the following trick. Since ω^r is the same in both the LNRF and comoving frames, we have

$$d\omega^r = -\omega^r{}_t \wedge \omega^t - \omega^r{}_\theta \wedge \omega^\theta - \omega^r{}_\varphi \wedge \omega^\varphi$$
$$= -\omega^r{}_t \wedge \gamma(\omega^\tau + v\omega^\psi) - \omega^r{}_\theta \wedge \omega^\theta - \omega^r{}_\varphi \wedge \gamma(v\omega^\tau + \omega^\psi)$$

[from (4.2.7)]

$$= -\omega^r{}_\tau \wedge \omega^\tau - \omega^r{}_\theta \wedge \omega^\theta - \omega^r{}_\psi \wedge \omega^\psi$$

by definition, from which

$$\omega^r_\psi = \gamma(\omega^r{}_\varphi + v\omega^r{}_t) \, . \tag{4.2.10}$$

Writing $\omega^r_\varphi = \Gamma^r_{\varphi t}\omega^t + \Gamma^r_{\varphi r}\omega^r + \Gamma^r_{\varphi\theta}\omega^\theta + \Gamma^r_{\varphi\varphi}\omega^\varphi$, and applying (4.2.7) again,

$$\Gamma^r_{\psi\tau} = \gamma^2(\Gamma^r_{\varphi t} + v\Gamma^r_{\varphi\varphi} + v\Gamma^r_{tt} + v^2\Gamma^r_{t\varphi}) \, , \tag{4.2.11}$$

and similarly

$$\Gamma^\theta_{\psi\tau} = \gamma^2(\Gamma^\theta_{\varphi t} + v\Gamma^\theta_{\varphi\varphi} + v\Gamma^\theta_{tt} + v^2\Gamma^\theta_{t\varphi}) \, . \tag{4.2.12}$$

Since e_t and e_φ commute, $\Gamma^r_{\varphi t} = \Gamma^r_{t\varphi}$, and by symmetry,

$$\Gamma^\theta_{\varphi t} = \Gamma^\theta_{\varphi\varphi} = \Gamma^\theta_{rt} = \Gamma^\theta_{r\varphi}$$

in the equatorial plane. Now either by direct calculation in the locally non-rotating frame, or from Bardeen et al. [36],

$$\Gamma^r_{\varphi t} = (\mu\beta/2\alpha)\,\omega' \, , \qquad \Gamma^r_{\varphi\varphi} = -\mu\beta'/\beta \, , \qquad \Gamma^r_{tt} = \mu\alpha'/\alpha \, ,$$

so that

$$\Gamma^\psi_{\theta\tau} = \Gamma^\theta_{rt} = 0 \, , \qquad \Gamma^r_{\psi r} =: v \, , \tag{4.2.13}$$

with

$$v = \gamma^2\mu[(1 + v^2)\,(\beta/2\alpha)\,\omega' + v(\alpha'/\alpha - \beta'/\beta)] \, . \tag{4.2.14}$$

Equation (4.2.9) now splits up to give

$$De_r = -ve_\psi \, , \qquad De_\psi = ve_r \, , \qquad De_\theta = 0 \, . \tag{4.2.15}$$

The significance of v is the following. At each point on the geodesic, there are three preferred spacelike directions e_r, e_θ, e_ψ. If an observer sets up a triad e_1, e_2, e_3, which coincides at one instant with e_r, e_θ, e_ψ, and is then parallely propagated along the geodesic, then the e_1, e_2, e_3 triad precesses about $e_2 = e_\theta$ with angular velocity $-v$. This kinematical effect is absent in Newtonian physics, but is well known in special relativity as the *Thomas precession* (see, e.g. [43]). From an orbiting satellite in a hybrid (Newtonian gravity + special relativity) theory, we would find a kinematical precession

$$v = (\gamma - 1)\, \Omega = (\gamma - 1)\, (GM/r^3)^{1/2}\,.$$

In the Kerr spacetime, however, Eq. (4.2.14) gives for our circular equatorial orbits

$$v = (GM/r^3)^{1/2}\,, \tag{4.2.16}$$

with no explicit a-dependence. We interpret this as meaning that both the e_1, e_2, e_3 and e_r, e_θ, e_ψ triads experience the same "frame dragging", so that their difference should not show this (a-dependent) effect.

We now write (4.2.8) as

$$e_{\alpha i} D^2 \eta^i = R_{ijkl} e_\alpha^i u^j (x^\beta e_\beta^k)\, u^l$$

$$= S_{\alpha\beta} x^\beta\,, \tag{4.2.17}$$

where

$$S_{\alpha\beta} = S_{\beta\alpha} = R_{ijkl} e_\alpha^i u^j e_\beta^k u^l\,. \tag{4.2.18}$$

$S_{\alpha\beta}$ has the following interpretation: consider a connecting vector with tetrad components x^α. If the (tidal) acceleration of this vector has tetrad components a_β, then

$$a_\beta = S_{\alpha\beta} x^\beta\,.$$

Since $S_{\alpha\beta}$ is symmetric we see that it is very similar to the stress tensor in continuum mechanics. Indeed if, as in the next section, we take x^α to be the coordinates of a point in a pressureless fluid, the analogy becomes closer. Because $R^{\tau\tau} = 0$ in a vacuum spacetime (R^{ab} is the Ricci tensor), $S_\alpha^\alpha = 0$, and so the effect of tidal accelerations is to produce a volume-preserving shear in the fluid.

To evaluate $S_{\alpha\beta}$ we use the techniques of the previous section. The NP tetrad in the equatorial plane is (as discussed in Section 2.1),

$$l^i = (2r^2)^{-1}\,(r^2 + a^2, \Delta, 0, a)$$

$$n^i = \Delta^{-1}(r^2 + a^2, -\Delta, 0, a)$$

$$m^i = (2r^2)^{-1/2}\,(ia, 0, 1, i)\,,$$

and we expand [16]

$$e_a = A_a l + B_a n + \bar{C}_a m + C_a \bar{m}\,. \tag{4.2.19}$$

---

[16] The A's, B's, and C's occurring here are analogous to the A, X, X'; B, Y, Y', and C, Z, Z' of Section 4.1.

Noting that $l - (\Delta r^2/2) n$ lies along e_r and $m + \bar{m}$ along e_θ, we immediately deduce

$$C_r = 0, \quad C_\psi = -\bar{C}_\psi, \quad C_\tau = -\bar{C}_\tau$$
$$C_\theta = \bar{C}_\theta = 2^{-1/2}, \quad A_\theta = B_\theta = 0. \tag{4.2.20}$$

Substituting (4.2.20) into (4.1.6) for a type D, $\{22\}$ vacuum spacetime gives, using the orthogonality of the tetrad,

$$S_{\alpha\beta} = 2\Psi_2 \begin{pmatrix} 1 - 3C_\tau^2 & 0 & 0 \\ 0 & -(1 - 6C_\tau^2)/2 & 0 \\ 0 & 0 & 1 - 3C_\tau^2 + 3C_\psi^2 \end{pmatrix}. \tag{4.2.21}$$

From the definition (4.2.18), $S_\alpha^\alpha = 0$ in vacuum, so

$$S_{\alpha\beta} = \Psi_2 \begin{pmatrix} 2(1 - 3C_\tau^2) & 0 & 0 \\ 0 & -(1 - 6C_\tau^2) & 0 \\ 0 & 0 & -1 \end{pmatrix}. \tag{4.2.22}$$

Noticing that $\Psi_2 = (M/r^3) = v^2$, Eqs. (4.2.15), (4.2.17), (4.2.22) combine to give [17]

$$\ddot{x}^r - 2v\dot{x}^\psi + 3v^2(2C_\tau^2 - 1) x^r = 0, \tag{4.2.23a}$$

$$\ddot{x}^\theta - 6v^2 C_\tau^2 x^\theta = 0, \tag{4.2.23b}$$

$$\ddot{x}^\psi + 2v\dot{x}^r = 0. \tag{4.2.23c}$$

This form of the geodesic deviation equation demonstrates that there will be no kinematical magnification of the tidal accelerations provided $C_\tau^2 = A_\tau B_\tau - 1/2 = O(1)$. For the geodesics under discussion, one computes

$$C_\tau^2 = -(\gamma^2/2\mathscr{A}) [a^2\Delta - 2av\Delta^{1/2}(r^2 + a^2) + v^2(r^2 + a^2)]. \tag{4.2.24}$$

It can be seen from Fig. 5 that $\gamma = O(1)$ for all stable circular orbits in the equatorial plane. We conclude that the relative tidal accelerations of neighbouring such orbits receives no kinematical magnification. It should be remarked however that this conclusion cannot be deduced directly from the theorem of the preceding section since, except in the case $a \sim M$ the tangent vector u^i is not close (in the sense of the theorem) to a RPND. Indeed, for $a < M$, bodies moving in unstable circular orbits close to the circular null geodesics (the r_γ curve in Fig. 3) will have large γ and do have their relative tidal accelerations magnified by the factor γ^2 (as one might have been lead to expect from Pirani's result [41]).

---

[17] A dot denotes differentiation with respect to proper time along the worldline.

4.3. The Roche Problem

In this section we present a simplified account of Fishbone's extension
of the above calculation to the Roche problem. Fishbone [42] considers
a self-gravitating gaseous body orbiting the black hole. The dimensions
of the body and internal velocities are supposed small, so that Newtonian
theory applies inside the body. If the internal pressure is also small, the
individual particles move almost along geodesics, and (4.2.23) gener-
alises to

$$\varrho Dv - 2\varrho \, v \wedge v = -Vp + \varrho V(\varphi_{BH} + \varphi_G). \tag{4.3.1}$$

Here v is the 3-velocity of the fluid $v = v e_\theta$, ϱ is the density of the fluid,
φ_G the self-gravitational field potential and φ_{BH} the potential due to
the black hole

$$\varphi_{BH} = B_{\alpha\beta} x^\alpha x^\beta := \tfrac{1}{2}(S_{\alpha\beta} + v^2 \, \delta_{\alpha\beta}) \, x^\alpha x^\beta. \tag{4.3.2}$$

It is possible to split $B_{\alpha\beta}$ uniquely into a trace-free part, the tidal forces
potential, and trace times $\tfrac{1}{3} \delta_{\alpha\beta}$ part, the centrifugal potential. We do
not need to do this here. The condition for Newtonian physics to be a
good approximation is

$$L \ll R, \tag{4.3.3}$$

where L is a typical dimension of the body, and R is the radius of curvature of spacetime. In cases where this is only marginally satisfied it
would be more accurate to use a post-Newtonian approximation.
Fishbone points out that the condition for equilibrium is, from (4.3.1),

$$p/\varrho = \varphi_{BH} + \varphi_G + \text{const}. \tag{4.3.4}$$

The fluid body will be in equilibrium if $\varphi_{BH} + \varphi_G$ is constant on isobaric
surfaces. Since φ_{BH} is quadratic in x^α, ellipsoidal figures are possible
with [18]

$$\varphi_G = -G\varrho \left(A + \sum_\alpha A_\alpha (x^\alpha)^2 \right), \tag{4.3.5}$$

where A, A_α are constants depending on the shape of the ellipsoid [77].
It is clearly necessary for equilibrium that the pressure should decrease
outward towards the surface or

$$B_{\alpha\alpha} - G\varrho A_\alpha < 0 \quad \text{(no sum over } \alpha) \tag{4.3.6}$$

or

$$\varrho > (1/\pi G) \, B_{\alpha\alpha}/A_\alpha. \tag{4.3.7}$$

---

[18]   The A, A_α here bear no relation to those of Sections 4.1, 4.2.

Now as was shown in the previous section, $B_{\alpha\alpha} \sim \gamma^2 GM/r^3$, and introducing a dimensionless radius $x = c^2 r/GM$, we have

$$\varrho > (c^6/(\pi G^3 M_\odot^2))(\lambda\gamma^2/x^3)(M_\odot/M)^2$$
$$> 2 \times 10^{17}(M_\odot/M)^2 (\lambda\gamma^2/x^3) \,\text{gm cm}^{-3}, \tag{4.3.8}$$

where λ is a factor of order unity incorporating details of the Kerr geometry and the shape of the ellipsoid. Formula (4.3.8) displays the usual Roche phenomenon. There is a lower limit to the density, the Roche limit ϱ_R, below which no equilibrium configuration can occur,

$$\varrho_R \sim 2 \times 10^{17}(M_\odot/M)^2 (\lambda\gamma^2/x^3) \,\text{gm cm}^{-3}. \tag{4.3.9}$$

Fishbone made a much more detailed analysis, explicitly evaluating the elliptic integrals implicit in Eq. (4.3.5). An example of his results is given in Fig. 6. Here he computes the Roche limit for the closest stable circular orbit. As explained in Section 3.4, the radius of this orbit decreases as a increases, with $x = 9$ for $a = -M$ (i.e. retrograde orbits) through $x = 6$ (Schwarzschild) to $x = 1$ ($a = M$). However the γ for this

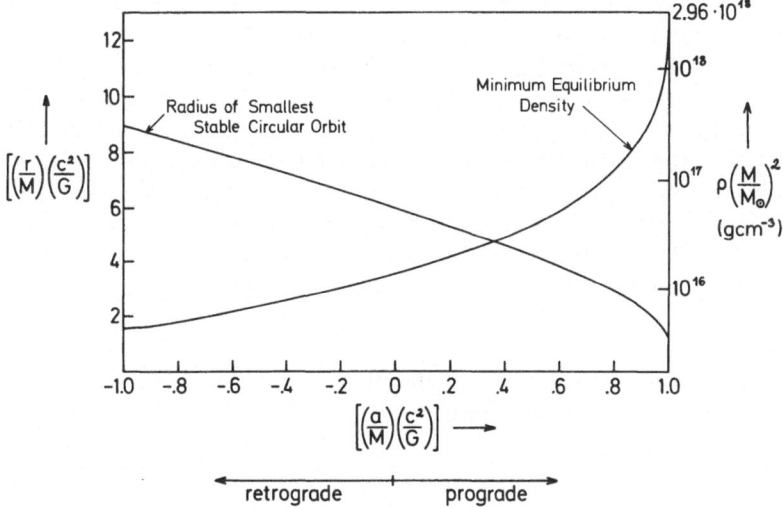

Fig. 6. *The Roche Limit for the Smallest Stable Orbit.* In this diagram the independent variable is the Kerr parameter a ($a > 0$ for prograde orbits, $a < 0$ for retrograde orbits). The radius of the smallest circular stable orbit in the equatorial plane is shown. This information is also contained in Fig. 3. Also shown is the Roche limit, the minimum density which a homogeneous fluid mass must have in order to stay in equilibrium. [Actually $\varrho(M/M_\odot)^2$ is plotted, where M is the mass of the black hole.] As $a \to Mc^2/G$, the smallest circular orbit moves closer to the black hole, so that the tidal forces increase. The Roche limiting density therefore also increases. This figure is adapted from [42]

orbit is $O(1)$ for all a. Thus as a ranges from $-M$ to M we expect a smooth increase of ϱ_R as is shown in Fig. 6. Since the photon orbit radius decreases as a increases, ranging from $x = 4$ $(a = -M)$ through $x = 3$ $(a = 0)$, to $x = 1$ $(a = M)$, we expect qualitatively similar behaviour for unstable high γ orbits close to the photon radius. Note however that in this case ϱ_R has to be multiplied by γ^2. Numerical examples are given in the next section.

4.4. Astrophysical Examples

We now consider two examples based on Fishbone's analysis. For our Sun to move in a stable orbit around a black hole

$$\varrho_R < \varrho_{\text{Sun}} \sim 1.4 \text{ gm cm}^{-3}.$$

This gives a *minimum* mass for the black hole $(M/M_\odot)_{\text{min}}$ as a function of a, with

$$5.4 \times 10^7 \leqq (M/M_\odot)_{\text{min}} \leqq 1.5 \times 10^9, \tag{4.4.1}$$

the lower limit corresponding to $a = -M$, the upper to $a = +M$. (Note that (4.4.1) does not assert $M/M_\odot \lesssim 1.5 \times 10^9$.) Now clearly within the sun, Newtonian physics applies, so that Fishbone's discussion is valid. Also if m is the mass of the orbiting body, $m/M \ll 1$ so that the test particle approximation also applies. If the sun were moving in an unstable high γ orbit, then $(M/M_\odot)_{\text{min}}$ would have to be *increased* by a factor γ. This illustrates the apparent paradox. For a given orbit, (constant x), the tidal forces decrease as M increases, due to the factor $(M/r^3) = x^{-3} M^{-2}$.

One can also make a crude analysis for a binary star system orbiting around a black hole. If the stars both have mass m, and separation d, then the gravitational acceleration on the stars due to each other is

$$F_{\text{Bin}} \sim Gm/d^2.$$

The relative acceleration due to the black hole is $\sim |S_{\alpha\beta}| d \sim GM\gamma^2 d/r^3$. If the stars are to remain bound

$$m/M \gtrsim \gamma^2 (d/r)^3,$$

or

$$(m/M_\odot)(M/M_\odot)^2 \gtrsim (d/(x R_{\text{Schw}}))^3 \gamma^2, \tag{4.4.2}$$

where R_{Schw} is the Schwarzschild radius of the sun (~ 1.5 km). For a typical binary system

$$M/M_\odot \gtrsim 10^{10} \gamma, \tag{4.4.3}$$

showing that binary systems are inherently less stable than single stars.

4.5. Gravitational Synchrotron Radiation

It is not intended to discuss gravitational synchrotron radiation (GSR) in any great detail here, since it is known that far more extensive reviews of this topic are in preparation. GSR nevertheless provided the initial impetus for the tidal forces calculations, and it is natural to discuss it in this context. Misner [44] originally hoped to explain Weber's gravitational wave observations [45] by considering bodies orbiting in the equatorial plane of a Kerr black hole in the center of our Galaxy at high γ. For *scalar* synchrotron radiation one would expect [44a] intense radiation to be emitted by the body in a narrow forward cone with semi angle

$$\Delta\theta \sim \gamma^{-1} \tag{4.5.1}$$

at all frequencies up to

$$\omega_{cutoff} \sim \gamma^2 \Omega, \tag{4.5.2}$$

where $\Omega = d\varphi/dt$ is the orbital angular velocity of the particle.

Since we subtend an angle 10^{-3} at the centre of the galaxy we deduce

$$\gamma \gtrsim 10^3. \tag{4.5.3}$$

If the cut off frequency is $\gtrsim 10^4 \sec^{-1}$, consonant with Weber's observations, Eq. (4.5.2) tells us nothing about Ω. However in the limiting case,

$$\Omega = 10^{-2}, \tag{4.5.4}$$

and since $\Omega \sim GM/r^3$, setting $r \sim M$ gives

$$M \sim 10^7 \, M_\odot. \tag{4.5.5}$$

Indeed this value is suggested by other astrophysical phenomena [25].

Now from (4.3.9),

$$\varrho_R = 2 \times 10^{17} \, (M_\odot/M)^2 \, (\lambda\gamma^2/x^3) \, \text{gm cm}^{-3}$$
$$\gtrsim 2 \times 10^3 \, \gamma^2 \, \text{gm cm}^{-3}. \tag{4.5.6}$$

In the stable, astrophysically plausible orbits we have $\gamma \sim 1$ so that the objects could be ordinary stars. However since $\gamma \sim 1$, no radiation would be emitted [36]. In unstable high energy orbits with $\gamma \sim 10^3$, one has $\varrho_R \sim 10^9 \, \text{gm cm}^{-3}$, so only white dwarfs and neutron stars would fail to be torn apart. Thus the tidal force calculation enables conditions to be placed on scalar GSR mechanisms although in this case the instability of the orbits itself renders the process astrophysically implausible. However it seems clear [46] that GSR can never produce the beaming effect for quadrupole radiation.

5. Perturbations of the Kerr Field

5.1. Introduction

In Section 1 the region outside a black hole was described by a stationary, asymptotically flat, vacuum spacetime with a regular event horizon, and it was conjectured that only the Kerr family of spacetimes with $a \leq M$ would satisfy these criteria. Clearly the Kerr spacetime would be of limited physical interest, however, if it were unstable in the sense that some initially small perturbation failed to remain small. It is therefore of prime importance to study the effect of small perturbations on Kerr's spacetime. No definite results are as yet known [50], so this section is necessarily of a pedagogical nature.

At first sight the study of stability seems straightforward; one introduces perturbed fields of amplitude $O(\varepsilon)$, where ε is some small real parameter, and linearizes the appropriate field equations by ignoring all terms which are of order $o(\varepsilon)$. The original spacetime is then stable provided the solutions of these equations remain "small". Since the perturbation equations are linear, they ought to be much easier to solve than the exact, usually nonlinear, field equations. However, a simple example will remind one of the difficulties involved in the interpretation of perturbation calculations within a covariant field theory. Consider the usual weak gravitational theory [79] based on a linearization about flat spacetime. If η_{ij} is the background Minkowski metric in some coordinate system, one writes the components of the metric tensor of the perturbed spacetime as

$$g_{ij} = \eta_{ij} + \varepsilon h_{ij}. \qquad (5.1.1)$$

Now one is permitted to make arbitrary infinitesimal coordinate transformations, usually called *gauge transformations*, in the perturbed spacetime of the form [19]

$$x^i \to x^i + \varepsilon \xi^i, \qquad (5.1.2)$$

and it is convenient to choose ξ^i to set $\partial h_i^j / \partial x^j = 0$. Gauge transformations such as this often simplify the calculations considerably. But there is a danger, because the change in coordinates was of the same order as the small quantities being calculated. If Q_0 is some observable quantity which becomes perturbed to $Q_0 + \varepsilon Q_1$, then one must be sure that under a gauge transformation

$$Q_0 + \varepsilon Q_1 \to Q_0 + \varepsilon Q_1 + o(\varepsilon). \qquad (5.1.3)$$

If (5.1.3) is satisfied, then Q_1, is said to be coordinate *gauge invariant*.

---

[19] ξ^i here is an arbitrary vectorfield.

In the case of weak gravitation, it is easy to check that the curvature tensor is coordinate gauge invariant. A less trivial example is geometrical optics on an arbitrary background spacetime [78]. It would clearly be a great advantage to be able to determine in advance whether the quantities one seeks are gauge invariant or not.

When explicitly writing down perturbation equations, it is often convenient to make use of a tetrad formalism [75], and to derive equations for the tetrad components of the perturbed fields, which are then scalars. This has the great advantage that the magnitude, or "smallness" of the perturbed quantities is thus unambiguously defined, which would not be the case for tensors. However, this approach introduce a new type of gauge transformation associated with infinitesimal transformation of the tetrad. Hence even if one works with tetrad components of the perturbed fields which happen to be coordinate gauge (CG) invariant, they will not in general be tetrad gauge (TG) invariant.

In the remaining sections, we study the question of the two kinds of gauge invariance, and find that

(i) one can make a coordinate and tetrad gauge invariant analysis of certain tetrad components of perturbed electromagnetic and gravitational fields in the NP formalism, and

(ii) the perturbation equations for these components can be decoupled from the remaining field components,

if and only if the unperturbed spacetime is vacuum (not electrovac), of algebraic type D {22}, and the NP tetrad is chosen with l^i and n^i lying along the RPND's.

Of course, the Kerr spacetime, with the tetrad given in Section 2.2, has precisely these properties, and the perturbation equations are readily obtained. Remarkably, it turns out that the decoupled equations are actually separable in Boyer-Lindquist coordinates, so that one has to deal only with ordinary differential equations. This was shown for scalar, spin zero, perturbations by Carter [47], and these equations can be solved [16]. Decoupled equations for the vacuum parts of electromagnetic, spin one [17, 48] and gravitational, spin two [18] perturbations have been obtained, and these have been separated by Teukolsky [18]. It is probable that the separability of these equations, like the decoupling, is a consequence of the algebraic type of the Kerr spacetime, being related to the Killing tensor (cf. Section 3.1), but the relation is as yet unclear.

Numerical integration of the decoupled, separated equations is in hand [50], and this should decide the stability of Kerr's spacetime. An indication of possible stability is already available, however. It is known in the Newtonian analysis of axisymmetric pulsations of a uniformly rotating, self gravitating gas [77] that instability always arises via a neutral mode of zero frequency. Chandrasekhar and Friedman [51] have found that no such modes exist for the Kerr spacetime, thus confirming a theorem of Carter [23, 24].

5.2. Gauge Invariance

A fuller account of the matters outlined in this and the following section will be published elsewhere [80]. Since actual perturbation analyses are usually carried out in a particular coordinate system, we discuss perturbations in coordinate terms and remark on the corresponding coordinate gauge transformations, making use of a generalisation of a result due to Sachs [76]. As mentioned in the introduction 5.1, it is often convenient to introduce a tetrad system to derive the perturbation equations, and in the second half of this section, we discuss the problem of tetrad gauge transformations.

Consider a spacetime manifold M with metric g. A perturbation of (M, g) is a 1-parameter family P of spacetimes $(M_\varepsilon, g(\varepsilon))$, which depends smoothly on the parameter ε, such that $M_0 = M$, $g(0) = g$. It is not part of the structure of P to say when a point of M_ε is "the same" as a point of the unperturbed spacetime M. In practice, however, one often wants to be able to answer this question since most perturbation analyses are carried out in terms of coordinate systems, and one wants to be able to relate coordinates on the M_ε to coordinates on M. Such a correspondence amounts to choosing a vectorfield X on P which is nowhere tangent to the M_ε; then all points along a flow line of X are "the same" as the point where this flow line meets M. Having selected an X, one has immediately introduced gauge transformations since any other vectorfield would have done just as well: the map relating two such choices is called a coordinate gauge (CG) transformation. This map will be a 1-parameter group of diffeomorphisms $\chi_\varepsilon: M_\varepsilon \to M_\varepsilon$ such that χ_0 is the identity transformation on M. Any perturbed quantity Q may be written

$$Q = Q_0 + \varepsilon Q_1 + o(\varepsilon) \tag{5.2.1}$$

where Q_0 is the unperturbed quantity and Q_1 (or more precisely εQ_1) is the perturbation. If this equation is invariant under CG transformations, i.e.

$$Q_1 \to Q_1 + o(1) \quad \text{as} \quad \varepsilon \to 0$$

then Q is said to be CG *invariant*. However it can be shown that under a CG transformation,

$$Q_1 \to Q_1 + \mathscr{L}_Y Q_0 + O(\varepsilon) \tag{5.2.2}$$

where $Y := (d/d\varepsilon)(\chi_\varepsilon)|_{\varepsilon=0}$, a vectorfield on M. Sachs has shown this for the metric, curvature tensor and its covariant derivatives, and contractions of these, but in fact the result holds for arbitrary scalars, vectors or tensors [80].

Thus Q_1 is CG invariant if and only if $\mathscr{L}_Y Q_0 = o(1)$, or $\mathscr{L}_Y Q_0 = 0$, since Q_0 and Y are independent of ε. But Y was completely arbitrary, which tells us that the only CG invariant quantities are those which are constant scalars or else vanish in the unperturbed spacetime.

This result can be obtained intuitively as follows. We can consider (M, g) to evolve to $(M_\varepsilon, g(\varepsilon))$ as a "time" coordinate ε increases. The maps associated with a given X above cause us to identify points in the two spaces M and M_ε which have zero Eulerian displacement. The χ_ε maps can be regarded as Lagrangian displacements with flow vector Y. Now it is well-known [77] that the equations for Eulerian and Lagrangian perturbations are identical if and only if either the flow is static, $Y = 0$, or the unperturbed quantities are zero.

The only tensors which can be constructed from the metric alone are the curvature tensor and its covariant derivatives, plus contractions of these. Since none of these quantities will vanish in a general spacetime, Sachs' lemma seems to indicate that a CG invariant perturbation analysis is generally impossible. The fact that the lemma holds more generally, however, points out another possibility: the Weyl tensor in an algebraically general spacetime defines four distinct future null directions at a generic point, and if the null vectors l^i and n^i in an NP tetrad are chosen to lie along two of these, then the curvature components Ψ_0 and Ψ_4 vanish: these two scalars are CG invariant in any perturbation analysis by the above results.

But the introduction of a tetrad introduces another type of freedom, namely tetrad transformations in the M_ε which reduce to the identity in the original spacetime. These are of two types, null rotations about l^i or n^i, and boosts in the $l^i - n^i$ plane combined with rotations in the orthogonal plane. The null directions of l^i and n^i are themselves invariant under the latter, so the best one can do is to present results in terms of quantities such as $\Psi_0 \Psi_4$ which are invariant under them. We confine our attention to infinitesimal null rotations, which we shall call tetrad gauge (TG) transformation.

For an infinitesimal null rotation about l^i we have

$$l^i \to l^i$$

$$m^i \to m^i + a l^i \tag{5.2.3}$$

$$n^i \to n^i + \bar{a} m^i + a \bar{m}^i + a\bar{a} l^i,$$

and under these

$$\Psi_0 \to \Psi_0$$

$$\Psi_4 \to \Psi_4 + 4\bar{a}\,\Psi_3 + 6\bar{a}^2\,\Psi_2 + 4\bar{a}^3\,\Psi_1 + \bar{a}^4\,\Psi_0\,.$$ (5.2.4)

Since (5.2.3) must reduce to the identity in the unperturbed spacetime, $a = O(\varepsilon)$. Hence Ψ_4 will be TG invariant if and only if $\Psi_3 = 0$ in the background. Similarly, Ψ_0 will be TG invariant if and only if $\Psi_1 = 0$ in the background. We therefore have the result that both Ψ_0 and Ψ_4 will be simultaneously CG and TG invariant if and only if the unperturbed Ψ_0, Ψ_1, Ψ_3 and Ψ_4 vanish. But this implies that l^i and n^i are RPND's, i.e. the unperturbed spacetime is of type $D\{22\}$.

Consider now the electromagnetic (em) case. If there is a non-null em field in (M, g), then l^i and n^i may be chosen to set $\varphi_0 = \varphi_2 = 0$. Under the transformations (5.2.3), however, $\varphi_2 \to \varphi_2 + 2\bar{a}\,\varphi_1 + \bar{a}^2\,\varphi_0$, which, since $\varphi_1 \neq 0$, is not TG invariant. In the case of a null em field, l^i can be chosen so that $\varphi_0 = \varphi_1 = 0$, so in this case φ_2 is TG invariant but not CG invariant. Hence a CG and TG invariant analysis of em perturbations can be made if and only if the em field vanishes in the unperturbed spacetime. This establishes the first part (i) of the result claimed in the introduction 5.1.

5.3. Decoupling of the Perturbation Equations

We have seen above that the algebraic type of the Kerr field, together with the properties of the associated NP tetrad, guarantees that one can perform a coordinate and tetrad gauge invariant analysis of the radiation parts φ_0, φ_2 of electromagnetic and Ψ_0, Ψ_4 of gravitational perturbations. In this section we write down perturbation equations for the analogous quantities in a general spacetime and then show that the algebraic type $D\{22\}$ of the Kerr field is in fact a necessary and sufficient condition that equations for these components decouple from the remaining tetrad field components. A more detailed derivation will be given elsewhere [80].

Use is made of a modified version of the NP formalism [29] given by Geroch, Held, and Penrose (GHP) [52]. This new formalism was designed as a computational tool for problems in which a pair of null directions is defined at each point of spacetime, and is therefore ideally suited to the Kerr field. It is useful also in general contexts, however, since an astonishing compactness of notation is achieved. Very briefly, the basic ideas are these: choose an NP tetrad with l^i and n^i lying along the two singled out null directions. The only freedom in choice of tetrad is then

$$l^i \to \lambda l^i$$
$$n^i \to \lambda^{-1} n^i \tag{5.3.1}$$
$$m^i \to e^{i\theta} m^i \, .$$

The first point about the GHP formalism is that it is given explicitly only in terms of quantities η which transform under (5.3.1) in the following simple way:

$$\eta \to \lambda^t \, e^{is\theta} \eta \, . \tag{5.3.2}$$

Such a quantity is said to be of boost weight t and spin weight s, or equivalently to be of type $(p, q) = \frac{1}{2}(t + s, t - s)$.

It turns out that all of the NP spin coefficients and field quantities are of well-defined type with the exception of the four spin coefficients α, β, γ and ε. A related fact is that the NP tetrad derivatives D, Δ, δ and $\bar{\delta}$, when applied to a quantity of well-defined type, do not in general result in quantity of well-defined type. The basic insight of GHP was that one can combine D, Δ, δ, $\bar{\delta}$ with α, β, γ, ε to define four new operators which do carry quantities of type to quantities of type, and that the non-type quantities then disappear from all equations. The new operators, acting on a quantity of type (p, q), are defined by

$$\text{\th} := D - p\varepsilon - q\bar{\varepsilon}$$
$$\text{\th}' := \Delta - p\gamma - q\bar{\gamma}$$
$$\eth := \delta - p\beta - q\bar{\alpha} \tag{5.3.3}$$
$$\eth' := \bar{\delta} - p\alpha - q\bar{\beta} \, .$$

Here, the symbol $\text{\th}$ is pronounced "thorn", and $\eth$ "edth".

The second point exploited by GHP is that the entire set of NP quantities and equations is transformed into itself by the following three formal operations on the tetrad:

$$\bar{l}^i = l^i, \bar{n}^i = n^i, \bar{m}^i = \bar{m}^i, \bar{\bar{m}}^i = m^i$$
$$l^{i'} = n^i, n^{i'} = l^i, m^{i'} = \bar{m}^i, \bar{m}^{i'} = m^i \tag{5.3.4}$$
$$l^{i*} = m^i, n^{i*} = -\bar{m}^i, m^{i*} = -l^i, \bar{m}^{i*} = n^i \, .$$

Consistent use of the $-$, $'$, and $*$ operations greatly reduces the required number of equations, all of the others following by applying (5.3.4) to the basic set. As a simple example of the notational compactness achieved, the four vacuum Maxwell equations follow by applying the transformations (5.3.4) to the simple equation

$$\text{Þ}\,\varphi_1 - \text{ð}\,\varphi_0 = \pi\,\varphi_0 + 2\varrho\,\varphi_1 - \varkappa\,\varphi_2\,.$$

Here and in all subsequent equations, we retain the NP notation for spin coefficients and field components, but use the GHP operators.

We will treat scalar (spin $s = 0$), em (spin $s = 1$), and gravitational (spin $s = 2$) perturbations of a vacuum spacetime. In the case of scalar perturbations, the coupling of the perturbed field to the spacetime via its stress energy and Einstein's equations first occurs at second order in the field, since the stress energy is quadratic in the field. Hence for first order perturbations one has only to solve the scalar wave equation

$$\square\,\chi = g^{ij}\nabla_i\nabla_j\chi = 0$$

in the unperturbed spacetime. Making use of the GHP formalism, this equation can be rewritten

$$(A_0 + A_0^{\bullet})\,\chi = 0\,, \tag{5.3.5}$$

where

$$A_0 := \text{Þ}'\text{Þ} + \bar{\mu}\text{Þ} - \varrho\text{Þ}'\,.$$

The electromagnetic case is slightly different since here, even though the field enters quadratically into its stress energy, the perturbation will be coupled to the spacetime via the Einstein-Maxwell equations to first order unless the unperturbed em field vanishes. Thus one can regard solutions of Maxwell's equations on the unperturbed background spacetime as first order em perturbations if and only if the unperturbed spacetime is a strict vacuum spacetime, not electrovac. To obtain an equation for φ_2, say, one differentiates one of the Maxwell equations and uses commutation relations to eliminate derivatives of φ_0. Further manipulation and use of field equations yields

$$(A_1 + A_1^{\bullet})\,\varphi_2 - 2[\text{Þ}\,\nu - \text{ð}\lambda - (\bar{\varrho} - 2\varrho)\,\nu - (\bar{\pi} + 2\tau)\,\lambda - 2\Psi_3]\,\varphi_1$$

$$+ [2\lambda\text{Þ}' - 2\nu\,\text{ð}' - 2(\pi\nu - \lambda\mu) - \Psi_4]\,\varphi_0 = 0\,, \tag{5.3.6}$$

where

$$A_1 := \text{Þ}'\text{Þ} + (2\mu + \bar{\mu})\,\text{Þ} - \varrho\text{Þ}' - 2\varrho\mu + \tfrac{1}{2}\Psi_2\,.$$

An analogous equation for φ_0 follows simply by priming (5.3.6).

Since the linearization required in any perturbation analysis has already been accomplished by using the scalar wave and Maxwell equations in the background, the curvature quantities, spin coefficients, and operators occurring in (5.3.5) and (5.3.6) are those of the unperturbed spacetime. The gravitational case is different. One proceeds in a formally similar manner to that employed in obtaining (5.3.6), using the Bianchi identities rather than Maxwell's equations, but now all of the quantities appearing are those of a new perturbed solution of Einstein's vacuum equations. The resulting equations must later be linearized about the particular spacetime in which we are interested. The result, for Ψ_4, can be written

$$(A_2 + A_2^*)\, \Psi_4 - 2[2\, \text{Þ}\, v - 2\partial\lambda - 2(\bar{\varrho} - 2\varrho)\, v - 2(\bar{\pi} + 2\tau)\, \lambda - 5\Psi_3]\, \Psi_3$$

$$+ [4\lambda\, \text{Þ}' - 4v\bar{\partial} - 12(\pi v - \lambda\mu) - 3\Psi_4]\, \Psi_2 = 0, \qquad (5.3.7)$$

where

$$A_2 := \text{Þ}'\text{Þ} + (4\mu + \bar{\mu})\, \text{Þ} - \varrho\, \text{Þ}' - 4\varrho\mu + \tfrac{1}{2}\Psi_2 .$$

Although we have not used a notational device to distinguish the unperturbed quantities in (5.3.5) and (5.3.6) from the perturbed quantities in (5.3.7), we emphasize that (5.3.7) is an exact equation in the perturbed spacetime, and has not yet been linearized.

We first treat decoupling in the electromagnetic case. If the background, unperturbed spacetime is of type $D\{22\}$ with the tetrad adapted to the RPND's, then $\varkappa = \sigma = \lambda = v = \Psi_0 = \Psi_1 = \Psi_3 = \Psi_4 = 0$, [30], and the coefficients of φ_0 and φ_1 in (5.3.6) vanish. The result is an equation

$$(A_1 + A_1^*)\, \varphi_2 = 0 \qquad (5.3.8)$$

for φ_2 alone, with A_1 as given above, which is decoupled from φ_0 and φ_1. It is clear that one can obtain a decoupled equation for φ_0 as well. Consequently, that the unperturbed spacetime be of type $D\{22\}$ is sufficient for the existence of decoupled equations for the "radiation parts" (defined by the RPND's) of an em perturbation. Conversely, the general behaviour of the spin coefficients and curvature [57] requires that a tetrad for which the coefficients of φ_0 and φ_1 in (5.3.6) vanish is one for which $\varkappa = \sigma = \lambda = v = \Psi_1 = \Psi_4 = 0$. The Goldberg-Sachs theorem [62] then implies that the unperturbed spacetime is of type $D\{22\}$, with the tetrad adapted to the RPND's. Again, (5.3.8) results.

The argument in the gravitational case is formally similar, although here we only require that the coefficients of the ψ's vanish to the order of the perturbation. Suppose that the unperturbed spacetime is of type $D\{22\}$, with the appropriate tetrad. Then the unperturbed values of $\varkappa$, σ, λ, v, Ψ_0, Ψ_1, Ψ_3, and Ψ_4 all vanish. These symbols in (5.3.7) do not vanish, however, but are of first order smallness. Linearisation simply consists of neglecting all terms in (5.3.7) which are products of these small quantities. Because of the type, and choice of tetrad in the unperturbed spacetime, the coefficient of Ψ_3 is of first order, and since Ψ_3 is itself of first order, this entire term can be neglected. Similarly, all of the terms occurring in the coefficient of Ψ_2, except $(3\Psi_4)\Psi_2$ are of second order, and can be neglected (one has to use the field equations to see this). Finally, the surviving first order term $3\Psi_4\Psi_2$ can be incorporated into the expression operating on Ψ_4 to give

$$(A_2 + A_2^* - 3\Psi_2)\Psi_4 = 0, \tag{5.3.9}$$

with A_2 as given above. Now Ψ_2, the operators, and the spin coefficients in (5.3.9) take their unperturbed values; the only perturbed quantity in (5.3.9) is Ψ_4. It is clear that one also obtains a decoupled equation for Ψ_0. Hence, type $D\{22\}$ is sufficient for one to obtain decoupled equations for Ψ_0 and Ψ_4. Conversely, suppose a tetrad has been chosen in the unperturbed spacetime so that $\Psi_4 = 0$. We see from (5.3.7) that, once the $3\Psi_4\Psi_2$ term has been incorporated into the expression $A_2 + A_2^*$ as above, a decoupled equation for the perturbed value of Ψ_4 can be obtained only if the remaining terms are of second order. This requires that Ψ_3 be of first order. Thus $\Psi_3 = 0$ in the background, from which v and λ also are of first order since the unperturbed spacetime is now algebraically special with n^i an RPND. That the surviving terms in (5.3.7) are also of second order then requires that $\varkappa = \sigma = 0$ in the background, so l^i is also an RPND [62]. It follows that the analogous equation for Ψ_0 decouples automatically.

We have established, in summary, the following result: for a vacuum spacetime any of the following implies the other two:

(i) there exists an NP tetrad in which the "radiative" parts (φ_0 and φ_1 in the em case; Ψ_0 and Ψ_4 in the gravitational) of the perturbed fields are coordinate and tetrad gauge invariant;

(ii) there exists an NP tetrad in which the perturbation equations for the radiative parts of the perturbed em and gravitational fields decouple from the rest of the perturbed field;

(iii) the unperturbed spacetime is of type $D\{22\}$ with NP tetrad adapted to the RPND's.

The above Eqs. (5.3.5), (5.3.8) and (5.3.9) can be combined into a master equation [54, 18]

$$(A_s + A_s^*)\chi_s = 0, \quad s = 0, 1, 2 \tag{5.3.10}$$

where

$$A_s := \text{Þ}'\text{Þ} + (2s\mu + \bar{\mu})\text{Þ} - \varrho\,\text{Þ}' - 2s\varrho\mu - s(s - 3/2)\Psi_2,$$

and

$$\chi_0 = \chi, \quad \chi_1 = \varphi_2, \quad \chi_2 = \Psi_4.$$

The analogous equation for the spin $s = +$ (spin weight) field components is

$$(B_s + B_s^*)\,\chi_{-s} = 0, \qquad s = 0, 1, 2 \tag{5.3.11}$$

where

$$B_s := \text{\th}'\,\text{\th} + \bar{\mu}\,\text{\th} - (2s+1)\,\varrho\,\text{\th}' - s(s+1/2)\,\Psi_2$$

and

$$\chi_0 = \chi, \qquad \chi_{-1} = \varphi_0, \qquad \chi_{-2} = \Psi_0.$$

We conclude with a remark on the separability of these equations, which has been demonstrated for $s = 0$ by Carter [47] and for $s = 1, 2$ by Teukolsky [18] when the equations are written out in BL coordinates. Held [53] has recently shown how one can formally integrate many of the GHP equations in a type {22} spacetime without having to introduce coordinates. It turns out that one can demonstrate the separability of (5.3.10) and (5.3.11) using Held's methods, and possibly even analyze the stability question itself. This work is in progress.

Acknowledgements. We are grateful to our colleagues Dieter Brill, Paul Chrzanowski, Jürgen Ehlers, Leslie Fishbone, Petr Hájícek, Peter Kafka and Bernd Schmidt for helpful discussions.

References

1. Kerr, R. P.: Phys. Rev. Letters **11**, 237 (1963).
2. Boyer, R. H., Lindquist, R. W.: J. Math. Phys. **8**, 265 (1967).
3. Kerr, R. P., Schild, A.: Applications of Nonlinear Partial Differential Equations, Proc. Symposia in Appl. Math. Vol. XVII, p. 199. ed. R. Finn, Providence: Am. Math. Soc. 1965.
4. Bardeen, J. M.: Astrophys. J. **162**, 71 (1970).
5. Penrose, R.: Contemporary Physics: Trieste Symposium 1968, Vol. I, p. 545. ed. A. Salam. Vienna: I.A.E.A. 1969.
6. Penrose, R.: Rivista Nuovo Cimento, Serie I, **1**, Numero Speciale, 252 (1969).
7. Ruffini, R., Wheeler, J. A.: The Significance of Space Research for Fundamental Physics, ed. A. F. Moore and V. Hardy. Paris: E.S.R.O. 1970.
8. Penrose, R., Floyd, R. M.: Nature Phys. Sci. **229**, 177 (1971).
9. Boyer, R. H., Price, T. G.: Proc. Cambridge Phil. Soc. **61**, 531 (1965).
10. Carter, B.: Phys. Rev. **174**, 1559 (1968).
11. de Felice, F.: Il Nuovo Cimento **57** B, 351 (1968).
12. de Felice, F., Calvani, M.: Univ. of Padova preprint, Nov. 1971.
13. Wilkins, D. C.: Phys. Rev. D **5**, 814 (1972).
14. Stewart, J. M., Walker, M.: Commun. Math. Phys. **29**, 43 (1973).
15. Fishbone, L. G.: Astrophys. J. **175**, L 155 (1972).
16. Brill, D. R., Chrzanowski, P. L., Martin Pereira, C., Fackerell, E. D., Ipser, J. R.: Phys. Rev. D **5**, 1913 (1972).

17. Fackerell, E. D., Ipser, J. R.: Phys. Rev. D **5**, 2455 (1972).
18. Teukolsky, S. A.: Phys. Rev. Letters **29**, 1114 (1972).
19. Penrose, R.: Phys. Rev. Letters **14**, 57 (1965).
20. Hawking, S. W., Penrose, R.: Proc. Roy. Soc. A **314**, 529 (1970).
21. Hawking, S. W.: Commun. Math. Phys. **25**, 152 (1972).
22. Penrose, R.: talk at Sixth Texas Symposium on Relativistic Astrophysics, New York, Dec. 1972.
23. Carter, B.: Phys. Rev. Letters **26**, 331 (1971).
24. Carter, B.: The Stationary Axisymmetric Black Hole Problem, Cambridge University preprint, 1972, Also: Black Holes, ed. C. de Witt and B. S. de Witt. New York: Gordon and Breach 1973.
25. Lynden-Bell, D.: Nature **223**, 690 (1969).
26. Bardeen, J. M.: Nature **226**, 64 (1970).
27. Carter, B.: Commun. Math. Phys. **17**, 233 (1970).
28. Carter, B.: J. Math. Phys. **10**, 70 (1969).
29. Newman, E. T., Penrose, R.: J. Math. Phys. **3**, 566 (1962).
30. Kinnersley, W. M.: J. Math. Phys. **10**, 1195 (1969).
31. Penrose, R.: Phys. Rev. Letters **10**, 66 (1963).
32. Penrose, R.: in Relativity, Groups, and Topology, eds. B. de Witt and C. de Witt. New York: Gordon and Breach 1964.
33. Penrose, R.: Proc. Roy. Soc. A **284**, 159 (1965).
33a. Bondi, H., van der Burg, M. G. J., Metzner, A. W. K.: Proc. Roy. Soc. A **269**, 21 (1962).
34. Winicour, J.: J. Math. Phys. **9**, 861 (1968).
35. Cohen, J. M.: J. Math. Phys. **9**, 905 (1968).
36. Bardeen, J. M., Press, W. H., Teukolsky, S. A.: Astrophys. J. **178**, 347 (1972).
37. Vishveshwara, C. V.: J. Math. Phys. **9**, 1319 (1968).
38. Walker, M., Penrose, R.: Commun. Math. Phys. **18**, 265 (1970).
39. Hughston, L. P., Penrose, R., Sommers, P., Walker, M.: Commun. Math. Phys. **27**, 303 (1972).
40. Geroch, R.: private communication (1969).
41. Pirani, F. A. E.: Proc. Roy. Soc. (Edinburgh) A **252**, 96 (1959).
42. Fishbone, L. G.: The Relativistic Roche Problem: Bodies in Equatorial, Circular Orbits Around Kerr Black Holes, University of Maryland, Department of Physics and Astronomy, Technical Report No. 73–025, Aug. 1972.
43. Fokker, A. D.: Time and Space, Weight and Inertia, p. 55. London: Pergamon Press 1965.
44. Misner, C. W.: Phys. Rev. Letters **28**, 994 (1972).
44a. Misner, C. W., Breuer, R. A., Brill, D. R., Chrzanowski, H. G., Hughes III, Pereira, C. M.: Phys. Rev. Letters **28**, 998 (1972).
45. Weber, J.: Phys. Rev. Letters **22**, 1320 (1969); **24**, 276 (1970); **25**, 180 (1970); Sci. Am. **224**, 22 (1971).
46. Ruffini, R.: private communication (1973); cf. also Black Holes, eds. C. de Witt and B. S. de Witt. New York: Gordon and Breach 1973.
47. Carter, B.: Commun. Math. Phys. **10**, 280 (1968).
48. Ipser, J. R.: Phys. Rev. Letters **27**, 529 (1971).
49. Press, W. H., Teukolsky, S. A.: Nature **238**, 211 (1972).
50. Press, W. H.: Black Hole Perturbations, Orange Aid Preprint, Jan. 1973.
51. Chandrasekhar, S., Friedman, J. L.: Astrophys. J. **177**, 745 (1972).
52. Geroch, R., Held, A., Penrose, R.: A spacetime calculus based on pairs of null directions, to appear in J. Math. Phys. **14**, (1973).
53. Held, A.: A formalism for the investigation of algebraically special metrics, preprint, 1972.

54. Bardeen, J. M., Press, W. H.: J. Math. Phys. **14**, 7 (1973).
55. Newman, E. T., Penrose, R.: J. Math. Phys. **7**, 863 (1966).
56. Newman, E. T., Penrose, R.: Proc. Roy. Soc. A **305**, 175 (1968).
57. Newman, E. T., Unti, T.: J. Math. Phys. **3**, 891 (1962).
58. Penrose, R.: Battelle Rencontres, eds. C. de Witt and J. Wheeler. New York: Benjamin 1968.
59. Boyer, R. H.: Proc. Roy. Soc. A **311**, 245 (1969).
60. Bardeen, J. M.: Astrophys. J. **161**, 103 (1970).
61. Floyd, R. M., Sheppee, B. A. V.: Int. J. Theor. Phys. **6**, 281 (1972).
62. Goldberg, J., Sachs, R.: Acta Phys. Polon. **22**, Suppl., 13 (1962).
63. Lynden-Bell, D., Rees, M. J.: Monthly Notices Roy. Astron. Soc. **152**, 461 (1971).
64. Christodoulou, D.: Phys. Rev. Letters **25**, 1596 (1970).
65. Geroch, R.: Ann. Phys. **48**, 526 (1968).
66. Cohen, J. M.: J. Math. Phys. **8**, 1477 (1967).
67. Florides, P. S.: Nuovo Cimento **13** B, 1 (1973).
68. Landau, L.: Phys. Z. Sowjetunion **1**, 285 (1934).
69. Chandrasekhar, S.: Monthly Notices Roy. Astron. Soc. **95**, 207 (1935).
70. Oppenheimer, J. R., Volkoff, G.: Phys. Rev. **55**, 374 (1939).
71. Harrison, B. K., Thorne, K. S., Wakano, M., Wheeler, J. A.: Gravitation Theory and Gravitational Collapse. Chicago: Univ. of Chigaco 1965.
72. Pajerski, D. W., Newman, E. T.: J. Math. Phys. **12**, 1929 (1972).
73. Hawking, S. W.: Black Holes, eds. C. de Witt and B. S. de Witt. New York: Gordon and Breach 1973.
74. Forster, J., Newman, E. T.: J. Math. Phys. **8**, 189 (1967).
75. Price, R. H.: Phys. Rev. D **5**, 2419, 2439 (1972).
76. Sachs, R. K.: in Relativity, Groups and Topology, p. 556, eds. B. de Witt, C. de Witt. New York: Gordon and Breach 1964,
77. Chandrasekhar, S.: Ellipsoidal Figures of Equilibrium. New Haven: Yale University 1969.
78. Isaacson, R. A.: Phys. Rev. **166**, 1263, 1272 (1968).
79. Sachs, R. K., Bergmann, P.: Phys. Rev. **112**, 674 (1958).
80. Stewart, J. M., Walker, M.: to be published.
81. Börner, G.: this volume.
82. Yodzis, P., Seifert, H.-J., Müller zum Hagen, H.: Hamburg Univerity preprint, 1973.
83. Kundt, W.: Proc. Can. Math. Congress, ed. R. Vanstone. Montreal: Can Math. Soc. 1972.

John Stewart
Martin Walker
Max-Planck-Institut für Physik und Astrophysik
D-8000 München 40
Föhringer Ring 6
Federal Republic of Germany

SPRINGER TRACTS IN MODERN PHYSICS

Ergebnisse der exakten Naturwissenschaften

Martin, B. R.: Kaon-Nucleon Interactions below 1 GeV/c (Vol. 55)
Morgan, D., Pišút, J.: Low Energy Pion-Pion Scattering (Vol. 55)
Oades, G. C.: Coulomb Corrections in the Analysis of πN Experimental Scattering Data (Vol. 55)
Pišút, J.: Analytic Extrapolations and the Determination of Pion-Pion Phase Shifts (Vol. 55)
Wanders, G.: Analyticity, Unitary and Crossing-Symmetry Constraints for Pion-Pion Partial Wave Amplitudes (Vol. 57)
Zinn-Justin, J.: Course on Padé Approximants (Vol. 57)

Regge Pole Theory, Dual Models

Ademollo, M.: Current Amplitudes in Dual Resonance Models (Vol. 59)
Chung-I Tan: High Energy Inclusive Processes (Vol. 60)
Collins, P. D. B.: How Important are Regge Cuts? (Vol. 60)
Collins, P. D. B., Gault, F. D.: The Eikonal Model for Regge Cuts in Pion-Nucleon Scattering (Vol. 63)
Collins, P. D. B., Squires, E. J.: Regge Poles in Particle Physics (Vol. 45)
Contogouris, A. P.: Certain Problems of Two-Body Reactions with Spin (Vol. 57)
Contogouris, A. P.: Regge Analysis and Dual Absorptive Model (Vol. 63)
Dietz, K.: Dual Quark Models (Vol. 60)
van Hove, L.: Theory of Strong Interactions of Elementary Particles in the GeV Region (Vol. 39)
Huang, K.: Deep Inelastic Hadronic Scattering in Dual-Resonance Model (Vol. 62)
Landshoff, P. V.: Duality in Deep Inelastic Electroproduction (Vol. 62)
Michael, C.: Regge Residues (Vol. 55)
Oehme, R.: Complex Angular Momentum (Vol. 57)
Oehme, R.: Duality and Regge Theory (Vol. 57)
Oehme, R.: Rising Cross-Sections (Vol. 61)
Rubinstein, H. R.: Duality for Real and Virtual Photons (Vol. 62)
Rubinstein, H. R.: Physical N-Pion Functions (Vol. 57)
Satz, H.: An Introduction to Dual Resonance Models in Multiparticle Physics (Vol. 57)
Schrempp-Otto, B., Schrempp, F.: Are Regge Cuts Still Worthwhile? (Vol. 61)
Squires, E. J.: Regge-Pole Phenomenology (Vol. 57)

Symmetries

Barut, A. O.: Dynamical Groups and their Currents. A Model for Strong Interactions (Vol. 50)
Ekstein, H.: Rigorous Symmetries of Elementary Particles (Vol. 37)
Gourdin, M.: Unitary Symmetry (Vol. 36)
Łopuszański, J. T.: Physical Symmetries in the

Framework of Quantum Field Theory (Vol. 52)
Pauli, W.: Continuous Groups in Quantum Mechanics (Vol. 37)
Racah, G.: Group Theory Spectroscopy (Vol. 37)
Rühl, W.: Application of Harmonic Analysis to Inelastic Electron-Proton Scattering (Vol. 57)
Wess, J.: Conformal Invariance and the Energy-Momentum Tensor (Vol. 60)
Wess, J.: Realisations of a Compact, Connected, Semisimple Lie Group (Vol. 50)

Weak Interactions

Barut, A. O.: On the S-Matrix Theory of Weak Interactions (Vol. 53)
Dosch, H. G.: The Decays of the $K_0 - \bar{K}_0$ System (Vol. 52)
Gasiorowicz, S.: A Survey of the Weak Interactions (Vol. 52)
Gatto, R.: Cabibbo Angle and $SU_2 \times SU_2$ Breaking (Vol. 53)
von Gehlen, G.: Weak Interactions at High Energies (Vol. 53)
Kabir, P. K.: Questions Raised by CP-Nonconservation (Vol. 52)
Kummer, W.: Relations for Semileptonic Weak Interactions Involving Photons (Vol. 52)
Müller, V. F.: Semileptonic Decays (Vol. 52)
Pietschmann, H.: Weak Interactions at Small Distances (Vol. 52)
Primakoff, H.: Weak Interactions in Nuclear Physics (Vol. 53)
Renner, B.: Current Algebra and Weak Interactions (Vol. 52)
Riazuddin: Radiative Corrections to Weak Decays Involving Leptons (Vol. 52)
Rothleitner, J.: Radiative Corrections to Weak Interactions (Vol. 52)
Segrè, G.: Unconventional Models of Weak Interactions (Vol. 52)
Stech, B.: Non Leptonic Decays (Vol. 52)

Nuclear Physics

Baryon-Baryon-Scattering

Kramer, G.: Nucleon-Nucleon Interactions Below 1 GeV/c (Vol. 55)
DeSwart, J. J., Nagels, M. M., Rijken, T. A., Verhoeven, P. A.: Hyperon-Nucleon Interactions (Vol. 60)

Electron Scattering

Theißen, H.: Spectroscopy of Light Nuclei by Low Energy (70 MeV) Inelastic Electron Scattering (Vol. 65)
Überall, H.: Electron Scattering, Photoexcitation and Nuclear Models (Vol. 49)

Quantum Statistics

Graham, R.: Statistical Theory of Instabilities in Stationary Nonequilibrium Systems with Applications to Lasers and Nonlinear Optics (Vol. 66)

Haake, F.: Statistical Treatment of Open Systems by Generalized Master Equations (Vol. 66)

Semiconductors

Feitknecht, J.: Silicon Carbide as a Semiconductor (Vol. 58)

Grosse, P.: Die Festkörpereigenschaften von Tellur (Vol. 48)

Schnakenberg, J.: Electron-Phonon Interaction and Boltzmann Equation in Narrow Band Semiconductors (Vol. 51)

Superconductivity

Lüders, G., Usadel, K.-D.: The Method of the Correlation Function in Superconductivity Theory (Vol. 56)

X-Ray, Neutron-, Electron-Scattering

Steeb, S.: Evaluation of Atomic Distribution in Liquid Metals and Alloys by Means of X-Ray, Neutron and Electron Diffraction (Vol. 47)

Springer, T.: Quasi-Elastic Scattering of Neutrons for the Investigation of Diffusive Motions in Solids and Liquids (Vol. 64)

To Appear in Forthcoming Volumes:

Überall, H.: Study of Nuclear Structure by Muon Capture

Levinger, J. S.: Two-Nucleon and Three-Nucleon Systems

Brandmüller, J., Claus, R.: Light Scattering on Optical Phonons and Polaritons

Langbein, D.: Theory of van der Waals Attraction